CHAOS, NOISE AND FRACTALS

Let's suppose that the first colour I apply to the canvas is red. Now that action determines everything else that happens to the painting. After that I could put yellow on it, and some blue, then I might perhaps obliterate the red with black and the blue would perhaps become yellow and the yellow purple, while the black changes to white. Obviously anything can happen. But the whole fascinating process began with that first red, and if I hadn't begun with red, the whole painting would have been different. Is there a system, is there order, is this chaos?

From an interview by Fréderic de Towarnicki in *Karel Appel*, Meulenhoff, Amsterdam 1984

MALVERN PHYSICS SERIES

General Series Editor: **Professor E R Pike** FRS

CHAOS, NOISE AND FRACTALS

Edited by

E R Pike
Royal Signals and Radar Establishment, Malvern
& Department of Physics, King's College, London
and
L A Lugiato
Dipartimento di Fisica, Politecnico de Torino

CRC Press
Taylor & Francis Group
Boca Raton London New York

CRC Press is an imprint of the
Taylor & Francis Group, an **informa** business

CRC Press
Taylor & Francis Group
6000 Broken Sound Parkway NW, Suite 300
Boca Raton, FL 33487-2742

First issued in paperback 2019

ISBN-13: 978-0-85274-364-5 (hbk)
ISBN-13: 978-0-367-40340-9 (pbk)

British Library Cataloguing in Publication Data

Chaos, noise and fractals.—(Malvern physics series; 3)
1. Chaotic behavior in systems. 2. Dynamics
3. Nonlinear theories
I. Pike, E.R. II. Lugiato, L.A.
515.3'5 QA845

Visit the Taylor & Francis Web site at
http://www.taylorandfrancis.com

and the CRC Press Web site at
http://www.crcpress.com

CONTENTS

PREFACE

The theory of non-linear dynamical systems has taken very much a second place to the development and refinement of that of linear systems over much of this century, in spite of a great deal of early pioneering work in the field by Poincaré, Birkhoff and others. A background level of research continued, as exemplified, for instance, by the work of Hopf in 1940 and later work by Krylov on mixing properties of ergodic systems, and the 'rich variety of behaviour, some of it very bizarre' found by Cartwright and Littlewood for the forced Van der Pol oscillator in 1943. However, it was not until the late 1950s and 1960s that the field really gathered momentum. In this period, work was typified by that of Kolmogorov, Arnold and Moser for Hamiltonian systems, that of Henon and Smale for more general diffeomorphisms, that of Faddeev, Marchenko, Kruskal and others in inverse scattering and, also, as the power of numerical computation increased dramatically, that of Lorenz in 1963 in the area of deterministic chaos in dissipative non-linear systems. These all led the way for very rapid recent developments, notably enhanced by universal features discovered in the routes to chaos.

The extensive use of Poincaré sections pointed out clearly the fractal nature of the strange attractors that underlie these chaotic motions, as also did various calculations of fractal dimension from numerical data. Related aspects are the interplay of deterministic chaos and stochastic noise, and the development of methods to distinguish them in experimental data. Another subject which has become a focus of attention in recent years is the rise of chaotic behaviour in quantum systems, and the features that characterise and limit the manifestations of chaos in quantum systems in comparison with classical systems.

This volume contains the invited contributions presented at a special seminar on the topics of Chaos, Noise and Fractals, generously funded by the London Branch of the US Government Office of Naval Research, to whom we express our gratitude on behalf of the participants. We also include two other contributions which were unable to be presented at the meeting because of the crowded programme, but whose authors were kind enough to provide manuscripts for these proceedings. The seminar was held at Villa Olmo, Como, Italy, 18 – 19 September 1986, immediately prior to a NATO Advanced Research Workshop on Quantum Chaos. The Proceedings of the Workshop will be published separately by Plenum Publishing Corporation, and the reader may benefit from considering the two volumes together.

We would like to thank Professor Guilio Casati and the staff of the Centre for Scientific Culture in Villa Olmo for their work in helping to organise the meeting. The participants enjoyed not only the presentations and the stimulating discussions in a friendly atmosphere, but also the magnificent location, which represents an ideal marriage between natural landscape and human architecture. We are grateful to Jane Zeuli and Gerda Wolzak for their enthusiastic and invaluable work during the two meetings, and also to Mrs Zeuli for considerable help with this publication afterwards.

It is a great pleasure to present this new volume in the Malvern Physics Series, and we express our appreciation to Adam Hilger, particularly Mr J Revill for their efficient assistance in its preparation.

E R Pike
L A Lugiato

January 1987

LIST OF CONTRIBUTORS

H ADACHIHARA Department of Mathematics
University of Arizona
Tucson
Arizona 85721
USA

F T ARECCHI Istituto Nazionale di Ottica
Largo E Fermi
6-50125 Firenze
Italy

M BRAMBILLA Dipartimento di Fisica
Università di Milano
Via Celoria 16
20133 Milano
Italy

D S BROOMHEAD Centre for Theoretical Studies
Royal Signals and Radar Establishment
St Andrews Road
Malvern
Worcestershire WR14 3PS
UK

H J CARMICHAEL Department of Physics
University of Arkansas
104 Physics Building
Fayetteville
Arkansas 72701
USA

G CASATI Dipartimento di Fisica
Università di Milano
Via Celoria 16
20133 Milano
Italy

B ECKHARDT Institut für Festkörperphysik
Kernforschungsanlage
D-5170 Jülich
FRG

M FEINGOLD
James Franck Institute
University of Chicago
5640 Ellis Avenue
Chicago
Illinois 60637
USA

H FRAHM
Institut für Theoretische Physik
Technische Universität Hannover
Appelstrasse 2
D-3000 Hannover
FRG

T GEISEL
Institut für Physik 1
Theoretische Physik
Universität Regensburg
Universität Strasse 31
Regensberg
FRG

R JONES
RSRE MOD(PE)
St Andrews Road
Malvern
Worcestershire WR14 3PS
UK

G P KING
Royal Signals and Radar Establishment
St Andrews Road
Malvern
Worcestershire WR14 3PS
UK

and

Department of Mathematics
Imperial College
Blackett Laboratory
London SW7 2BZ
UK

P L KNIGHT
Department of Physics
Blackett Laboratory
Imperial College
London SW7 2BZ
UK

L A LUGIATO Dipartimento di Fisica
Politecnico di Torino
c.so Duca Degli Abruzzi 24
10129 Torino
Italy

D W MCLAUGHLIN Department of Mathematics
University of Arizona
Tucson
Arizona 85721
USA

P MEYSTRE Optical Sciences Center
University of Arizona
Tucson
Arizona 85721
USA

H J MIKESKA Institut für Theoretische Physik
Technische Universität Hannover
Appelstrasse 2
D-3000 Hannover
FRG

J V MOLONEY Department of Physics
Heriot-Watt University
Riccarton
Edinburgh EH14 4AS
UK

L M NARDUCCI Department of Physics and Atmospheric Sciences
Drexel University
Philadelphia
Pennsylvania 19104
USA

A C NEWELL Department of Mathematics
University of Arizona
Tucson
Arizona 85721
USA

S J D PHOENIX Department of Physics
Blackett Laboratory
Imperial College
London SW7 2BZ
UK

E R PIKE

Department of Physics
King's College
Strand
London WC2R 2LS
UK

and

Centre for Theoretical Studies
Royal Signals and Radar Establishment
St Andrews Road
Malvern
Worcestershire WR14 3PS
UK

G RADONS

Institut für Physik 1
Theoretische Physik
Universität Regensburg
Universität Strasse 31
Regensberg
FRG

J RUBNER

Institut für Physik 1
Theoretische Physik
Universität Regensburg
Universität Strasse 31
Regensberg
FRG

SARBEN SARKAR

Centre for Theoretical Studies
Royal Signals and Radar Establishment
St Andrews Road
Malvern
Worcestershire WR14 3PS
UK

J S SATCHELL

Centre for Theoretical Studies
Royal Signals and Radar Establishment
St Andrews Road
Malvern
Worcestershire WR14 3PS
UK

and

Clarendon Laboratory
University of Oxford
Parks Road
Oxford OX1 3PU
UK

G STRINI Dipartimento di Fisica
Università di Milano
Via Celoria 16
20133 Milano
Italy

F VIVALDI Department of Mathematics
Queen Mary College
327 Mile End Road
London E1 4NS
UK

E M WRIGHT Optical Sciences Center
University of Arizona
Tucson
Arizona 85721
USA

HYPERCHAOS AND 1/*f* SPECTRA IN NONLINEAR DYNAMICS

F T ARECCHI

1. INTRODUCTION

In the middle 1300 the following problem has been attributed to Johannes Buridanus, a philosopher at the University of Paris. Suppose a donkey is just halfway between two equivalent choices (e.g. two food baskets that we call F1 and F2). What will be its decision? In the solution attributed to Buridanus the donkey dies, having no elements to decide for either solution. The current modern solution, upon which most of statistical physics is built, is more optimistic. The initial condition between the two choices is an unstable one, like the maximum x=0 in a quartic potential well $V(x)=-ax^2+bx^4$ ($a,b > 0$) and it would be left immediately once the donkey (taken as a material point initially at x=0) is coupled to the rest of the Universe, which provides for a thermal bath including fluctuations (even at zero temperature there would be quantum flunctuations).

Let us model the fluctuations as an additive white noise (no memory) source. If we use a discrete time approach and introduce an uncertainty Δx per step (the donkey's feet have a finite size), there is a single time scale τ, that corresponding to the average first passage time through Δx. Afterwards, because of the uniqueness theorem for the solution of a differential problem, noise will not play any extrarole and the donkey will go either to F1 or F2. The time scale τ provides an exponential decay of correlations, that is, of the memory of the initial uncertainty and an associated Lorentzian power spectrum

$$G(\omega) \simeq \frac{1/\tau}{\omega^2 + 1/\tau^2} \tag{1}$$

As well known a log-log of (1) has two asymptotic straight lines, a high frequency one with a slope -2 (20 db/decade) and a horizontal one for low frequency, corresponding to lack of correlations (white spectrum). The two lines cross at $\omega = 1/\tau$. The long time lack of correlation is the basis of all Markoffian approaches to statistical physics.

Buridanus' solution would be then wrong, since the donkey does not die but it performs a decision with a definite time scale.

These considerations where the basis of an approach to decay of unstable states motivated by an early experiment on a transient laser (Arecchi et al, 1967 and 1971) and then formalized in a general procedure (Arecchi et al, 1980 and 1982a).

If, however, still with the same two-valley potential (bistable solution) and in the presence of a white noise, we increase the number of degrees of freedom up to 3 in order to allow for a chaotic dynamics, then we observe experimentally, the possibility of jumps back and forth from a decision to the other one.

Aim of this paper is to show that this is equivalent to provide Buridanus'donkey with a fractal boundary between the two choices. Indeed an irregular rugged boundary can be crossed from several directions, and it will provide a large number of length scales rather than a single one Δx, and hence a large number of time scales. This will be equivalent to the superposition of many Lorentzians spectra as (1), thus providing a power spectrum utterly different from a Lorentzian one. The donkey will keep a long memory of the initial uncertainty and it might die as expected by Buridanus.

In the following we approach the problem with reference to chaotic dynamics, offering a solution in terms of an elementary model.

2. NOISE INDUCED TRAPPING AT THE BOUNDARY BETWEEN TWO ATTRACTORS

Addition of random noise in a nonlinear dynamical system with more than one attractor may lead to 1/f spectra, provided that the basin boundary be fractal (Arecchi and Califano, 1986). Combining the features leading to deterministic chaos with a random noise is somewhat equivalent to a double randomness and we call "hyperchaos" such a situation. Indeed random-random walks in ordinary space, as diffusion in disordered systems, have shown a 1/f behavior (Sinai, 1982; Marinari et al, 1983). Thus, hyperchaos here introduced is a random-random walk in phase space, where in fact one of the two sources of complex behavior is due to the fractal structure arising from deterministic dynamics.

To evaluate the impact of the following arguments, I premise some historical remarks on 1/f spectra in nonlinear dynamics.

Some years ago it was discovered (Arecchi and Lisi, 1982) that in a nonlinear dynamical system with more than one attractor, introduction of random noise induces a hopping between different basins of attraction, giving rise to a low frequency spectral divergence, resembling the 1/f noise well known in many areas of physics (Fig. 1). Such a discovery was confirmed by a laser experiment implying two coexisting

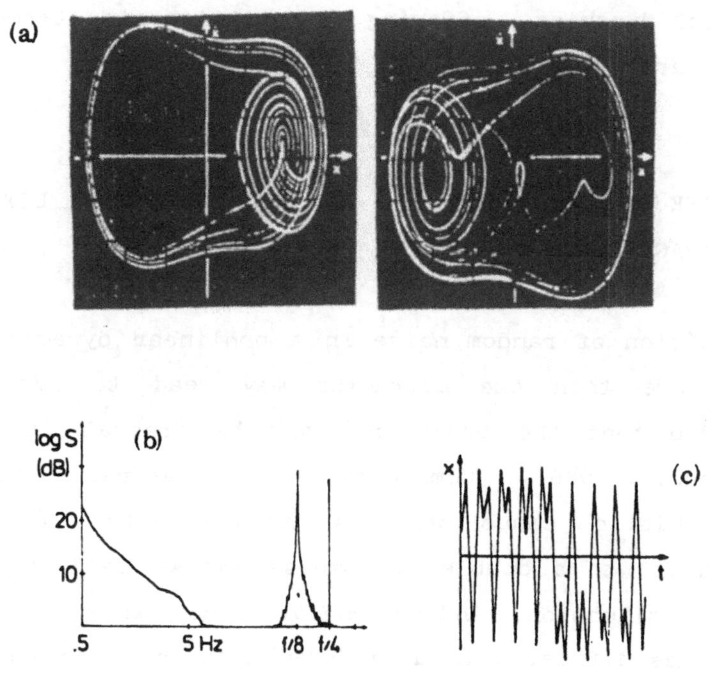

Figure 1
Electronic non linear forced oscillator obeying the law $\ddot{x} + \gamma \dot{x} - ax + bx^3 = A\cos(2\pi ft)$. Hopping between two attractors and associated 1/f spectrum in the purely bistable case. (a) Symmetric phase-space plots; (b) log-log spectrum showing the low frequency divergence, a broadened f/8 line, and a narrow f/4 line; (c) a sample of the x(t) plot.

attractors (Arecchi et al, 1982b) (Fig. 2), and later the effect was observed in other areas as e.g. Josephson tunnel junctions (Miracky et al, 1983).

The effect was questioned with two objections:

a) a noise induced jump across a boundary leads to a

telegraph signal, hence to a single Lorentzian spectrum (Beasley et al, 1983);

b) a computer experiment yielded a power law only over a limited spectral range (Voss, 1983).

The questions were answered (Arecchi and Lisi, 1983) with a statement of the empirical conditions under which the 1/f spectra appeared, namely:

i) coexistence of at least two attractors (so called " generalized multistability" (Arecchi et al, 1982b)),

ii) presence of noise,

iii) some "strangeness" in the attractors.

As a matter of fact this third condition was rather vague. To make it more precise, two theoretical models were explored, namely, a one dimensional cubic iteration map with noise (Arecchi et al, 1984a) and a forced Duffing equation with noise (Arecchi et al, 1984b; Arecchi et al, 1985a). Both these papers disclose interesting features, bringing more light on the above assumption iii). Fig. 2 of Arecchi et al 1984a shows that the size of the 1/f spectral region increases with the r.m.s. of the applied noise, that is, with the probability of crossing the basin boundary by a noise-induced jump.

The numerical evaluation of Arecchi et al 1984b and Arecchi et al 1985a showed that for some control parameters the boundary between basins of attraction was an intricated set of points, through which it was impossible to draw a simple line. In such cases the noise was most effective in yielding low frequency spectra 1/f-like.

On the other hand a fundamental logical approach to the 1/f problem was based on the composition of a large number of Lorentzians (or elementary Markov processes with exponential decay) whose weights are log-normally distributed (Montroll and Shlesinger, 1982), thus fulfilling the relation

$$\int_{\gamma_1}^{\gamma_2} \frac{\gamma}{\omega^2+\gamma^2}\, p\,(\gamma)d\gamma \simeq \text{const.x}\, \frac{1}{\omega}, \tag{2}$$

provided $p(\gamma) \sim 1/\gamma$, and for the frequency range $\gamma_1 \ll \omega \ll \gamma_2$.

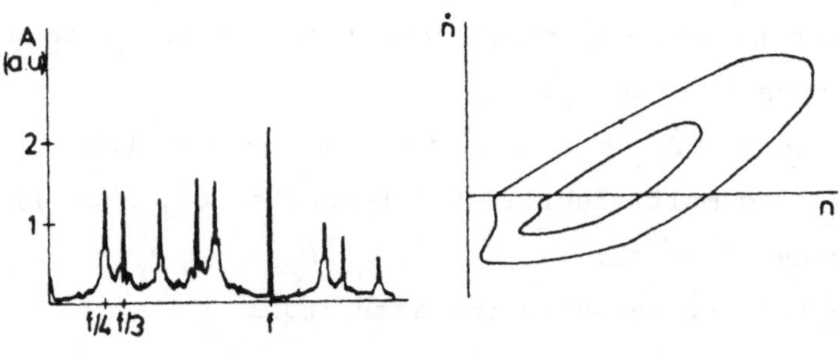

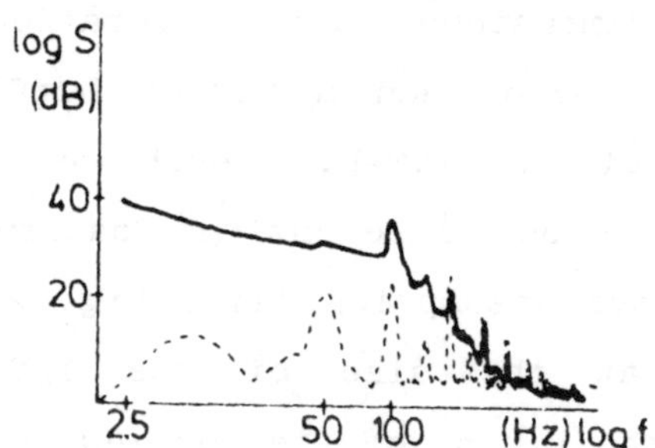

Figure 2
Bistability in a CO_2 laser with loss modulation. (a,b) coexistence of two attractors (period 3 and 4 respectively) high frequency spectrum around 100KHz, (c) comparison between the low frequency cut-off when the two attractors are stable (dashed line) and the low frequency divergence when noise is added (solid line).

Motivated by the rate processes considerations, which yielded a single Lorentzian for two attractors, we developed a kinetic model (Arecchi et al, 1984a) based on a single transition rate for each pair of attractors. In the case of M attractors, this yielded M-1 Lorentzians. To approximate the integral (2) by a sum (5% accuracy in fitting a 1/f law would require about one pole per decade) a large number $M >> 2$ of attractors is necessary and hence the integral of eq. (2) would be replaced by the sum over the M-1 Lorentzians corresponding to the eigenvalues of the kinetic model, however there is no reason to weigh the Lorentzians according to their reciprocal widths, hence no satisfactory reconstruction of a 1/f spectrum was possible. In fact, an experiment on a forced and noisy Duffing oscillator with an increasing number of attractors (Arecchi and Califano, 1984) did not offer a clear evidence of the expected scaling of the spectral exponent with the number of attractors. On the contrary, Arecchi et al 1984b and Arecchi et al 1985a showed that the boundary region between just two attractors was sufficient to yield 1/f-like spectra, at variance with the many-attractor model. Thus, this suggested that the boundary structure was the real responsible for a large number of decay constants (possibly log-normally distributed).

In the meantime, the fractal, structure of a basin boundary was explored in some examples (Grebogi et al, 1983; Mc Donald et al, 1985). This means the following. As the phase point wanders within one basin of attraction, if we draw a sphere around the point defining its distance from the other basin of attraction, the radii of these spheres are distributed with all scale lengths, according to the self similar structure of the fractal boundary.

Based on the above considerations, we have built an elementary cellular automaton which models the motion of the phase point within a fractal basin boundary under the presence of random noise. We model the boundary region of two basins of attraction A and B as two adjacent one-dimensional lattices of sites. Suppose we start from site i. At each discrete time step, if i belongs to A ($i = i_A$) it moves one step forward on the same lattice ($i_A \rightarrow (i_A + 1)$) and if it belongs to B it goes one step backward ($i_B \rightarrow (i_B - 1)$). In the absence of noise, once the motion has started on one basin, it will remain on it forever. In the presence of noise, at each time step there is a finite probability of a "cross" jump at the same lattice site, from stripe A to B: $i_A \rightarrow i_B$.

We call L the maximum size of the boundary region and $\ell_i \leq L$ any of the possible sizes of the fractal set. At each time step, the probabilities of permanence and jump are respectively

$$
\begin{aligned}
P_{AA} &= P_{BB} = \ell_i/L \\
P_{AB} &= P_{BA} = 1 - \ell_i/L
\end{aligned}
\qquad (3)
$$

To build a self-similar structure we allow ℓ_{ik} to scale as $\frac{\ell_{ik}}{L} = (1/2)^{V(i_k)}$ where $V(i_k)$ is a natural number sorted randomly for each site i_k ($i = -\infty$ to ∞, $k = A,B$). To deal with a real numerical experiment we consider finite sequences of N sites (e.g. $N = 10^3$) and we truncate the fractality by imposing $0 \leq V(i_k) < F$. Here, F is a finite integer denoting the maximum partitioning $(1/2)^{F-1}$, that is, the ultimate resolution of the measuring device in appreciating the fractal structure of our set. With all this in mind, for each evolution we extract a double sequence of N integers randomly distributed between 0 and F-1, and denote each site i_k by the corresponding number $V(i_k)$. This means that we have attributed to each site an "area of respect", that is, a specific separation ℓ_{i_k} from the other attractor, with ℓ_{i_k} depending on $V(i_k)$ as shown above. We start, e.g. on the basin A from $i_A = N/2$.

At this step, to account for a suitable noise yielding the permanence and jump probabilities (3), we generate a random number y uniformly distributed between 0 and 1. If

$y \lessgtr (1/2)^{V(i_A)}$, then at the next time the point goes to $i_A + 1$ on attractor A; if $y > (1/2)^{V(i_A)}$, then the point jumps instanteneously to site i_B and at the next time it goes to $i_B - 1$ on attractor B.

By measuring the position coordinate, taking the Fourier transform and squaring it, we can build the power spectra, that is, the transforms of the position correlation functions.

In Fig. 3 we show two power spectra for F=4, and 14 respectively. In fact, we have measured spectra for all integer values of F between 4 and 14, but we just report two samples over slightly more than three frequency decades. The sequence shows that, as the fractality increases, the slope of the log-log plot goes from about 2 (single Lorentzian) to about 1 (1/f spectrum). This appears better in Fig. 4, where the slope α of the $f^{-\alpha}$ spectral law is plotted versus the fractality F. The Lorentzian ($\alpha = 2$) of the random telegraph model is easily recovered for F=1, thus showing that noise induced jumps across a regular line boundary fulfill the intuitive expectation of a single decay rate. An analogy with the random-random walk (Sinai, 1984 and Marinari et al, 1983) is easily drawn. Indeed our motion is bound with an r.m.s. deviation going from about $\sqrt{t}$ to $|\log t|^2$ as the fractality F increases from 4 to 14, according to Sinai.

For comparison we mention other approaches leading to 1/f or anyway nonLorentzian low frequency spectra:

i) Pomeaeu-Manneville type-3 intermittency corresponds to slowly diverging trajectories with a 1/f power spectrum (Pomeau and Manneville, 1980; Procaccia and Schuster, 1983). This behavior is intrinsic to the dynamics, hence it occurs without noise.

ii) A deterministic diffusion process may occur beyond "crisis" (Geisel and Thomae, 1984; Geisel et al, 1985)

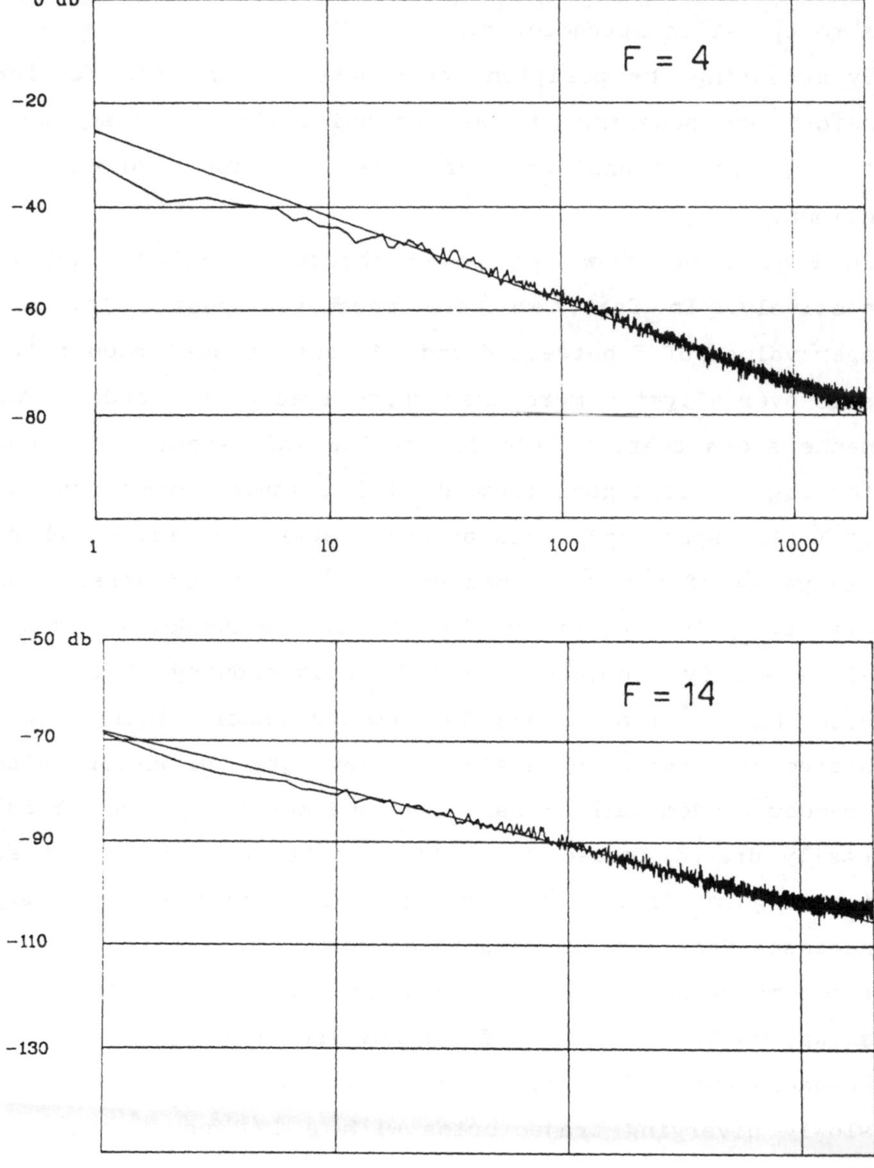

Figure 3
Power spectra (vertical) versus frequency (horizontal) in log-log scale. Wavy lines: measured spectra, straight lines: best fits, whose slopes α are reported in the next figure. The two samples shown refer to F = 4, and 14, respectively.

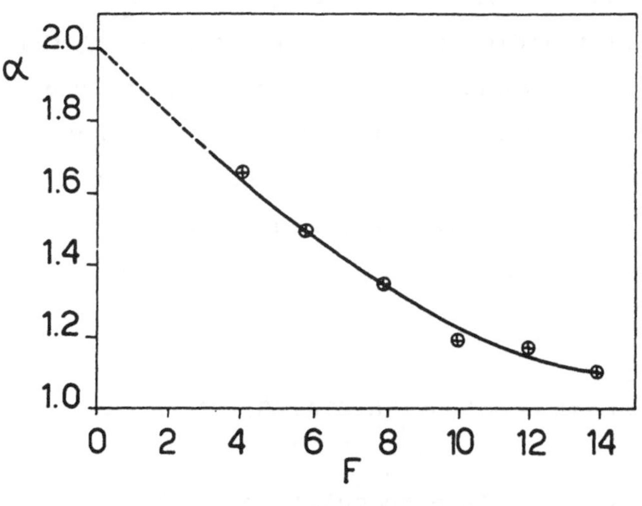

Figure 4

Exponents α of the power law $f^{-\alpha}$ versus fractality F.

when two otherwise disjoint attractors merge into a single one. Here again no noise is required, and a comparison of this behavior with noise induced jumps was given in Arecchi et al 1984a and Arecchi et al, 1985b.

iii) Another comparison of intrinsic versus noise -induced intermittency was carried on for a damped driven pendulum, which models a Josephson juction (Gwinn and Westervelt, 1985). This last papers offers numerical evaluations of spectra, showing an 1/f region extending over two decades, but to our knowledge nobody has tried so far to analyze the role of fractality and draw a comparison with Sinai subdiffusive motion.

Among other things, the results of this paper may strongly affect our current understanding of Optical Bistability (OB) phenomena. OB is described in terms of two fixed point attractors, which however are the result of a collective dynamics implying many degrees of freedom. There are no exhaustive analyses of the structure of the basin boundary, thus possible fractal structures may appear if the dynamics is evaluated in detail. On the other hand, in order to reduce the signal power necessary to drive the OB device from one state to the other, the system is usually set very near to the boundary. Thus, unavoidable random noise might induce low frequency spectra of the type above described.

3. CONCLUSION: LONG MEMORY IN STATISTICAL PHYSICS

Let me conclude with a speculation on the role of the long time terms in nonequilibrium statistical mechanics.

We have shown that, whenever in non linear dynamics more than one attractor is present, there are two distinct power spectra:

i) a high frequency one, corresponding to the decay of correlations within one attractor;

ii) a low frequency one, corresponding to noise induced jumps.

Based upon i), the usual transport coefficients for macroscopic equations of evolution have been built. Effect ii) has been overlooked so far. Here, I wish to consider an example showing the relevance of ii) with respect to multiphoton molecular excitation.

Let me consider a molecule with two isomeric states (cis and trans) of almost equal energy, separated by an energy barrier, say, of 1 eV (e.g. rodhopsin molecule in the retina of vertebrates).

We know that an IR laser such as a CO_2 laser (λ =10 μm, $h\nu \simeq 0.1$ eV) may give rise to a multiphoton absorption process if it is powerful enough to provide 10 photons within one coherence time of the "cis" valley, so that 10 small photons pile up to 1 eV excitation. (We are considering a molecule large enough so that the barrier is already a classical one, and so quantum tunneling is possible). We know that a vibrational mode IR active decays by intra-molecular relaxations toward the thermal bath of all other modes, in a time of order of 10^{-12}s =1 ps. In order to have a multiphoton isomeric transition, we should have a laser power of

$$10 \text{ photons/1ps} \simeq 10^{-7} \text{ watt}$$

over a cross-section of $\sim$ $(1\text{Å})^2$ 10^{-16}cm^2, and thus a laser intensity of 10^9 W/cm . But this was a Markovian point of view, based on a memory time of 1ps related to the high frequency spectral broadening. A double potential Valley dynamics is described by a Duffing equation (see Arecchi and Lisi, 1982; Arecchi eyt al, 1984b; Arecchi et al, 1985a) and the presence of an IR laser illumination as a forcing term yields a motion on an attractor not necessarily confined in one valley, even for very low intensities (see Fig. 1).

Such a chaotic motion may pass near to the boundary, hence requiring an activation energy much less than the barrier of 1 eV to be introduced into an Arrhenius type law. For instance, in Fig. 2 we have seen a high frequency spectrum around 100 KHz, and the corresponding low frequency jump spectrum at 1 Hz (5 decades below). By the same reasoning, we might expect that an intensity 5 or 6 decades lower (that is 10 photons/μs or just 10^3 W/cm) might be sufficient for a multiphoton isomerization process.

If we could use such a large enhancement factor in most activation processes of biochemical relevance, the consequence would be that the times necessary for biochemical evolution on Earth could be correspondingly reduced.

This is just a guess, to show how the introduction of the long memory processes here described for the first time may open new routes in the physics of complex systems.

REFERENCES

Arecchi, F.T., Degiorgio, D., and Querzola, B. 1967, Phys. Rev. Lett. 19, 1168

Arecchi, F.T., and Degiorgio, V. 1971, Phys. Rev. A3, 1108

Arecchi, F.T., and Politi, A. 1980, Phys. Rev. Lett. 45, 1215

Arecchi, F.T., Politi, A., and Ulivi, L. 1982, Nuovo Cimento 71B, 119

Arecchi, F.T., and Lisi, F. 1982, Phys. Rev. Lett. 49, 94

Arecchi, F.T, Meucci, R., Puccioni, G.P., and Tredicce, J.R. 1982, Phys. Rev. Lett. 49, 1217

Arecchi, F.T., and Lisi, F. 1983, Phys. Rev. Lett. 50, 1330

Arecchi, F.T., Badii, R., and Politi, A. 1984, Phys. Rev. A29, 1006

Arecchi, F.T., Badii, R., and Politi, A. 1984, Phys. Lett. A103, 3

Arecchi, F.T., and Califano, A. 1984, Phys. Lett. A101, 443

Arecchi, F.T., Badii, B., and Politi, A. 1985, Phys. Rev. A32, 402

Arecchi, F.T., and Califano, A. 1986, Europhysics Letters 2,

Beasley, M.R., D'Humieres, D., and Huberman, B.A. 1983, Phys. Rev. Lett. 50, 1328

Geisel, T., and Thomae, S. 1984, Phys. Rev. Lett. 52, 1936

Geisel, T., Nierwetberg, J., and Zachere, A., 1985, Phys. Rev. Lett. 54, 616

Grebogi, G., Ott, E., and Yorke, J.A. 1983, Phys. Rev. Lett. 50, 935

Gwinn, E.G., and Westervelt, R.M. 1985, Phys. Rev. Lett. 54, 1613

Marinari, E., Parisi, G., Ruelle, D., and Windey, P. 1983, Phys. Rev. Lett. 50, 1223

Mc Donald, S.M., Grebogi, C., Ott, E., and Yorke, J.A. 1985, Physica 17D, 125

Miracky, R.F., Clarke, J., and Koch, R.H. 1983, Phys. Rev. Letters 50, 856

Montroll, E.W., and Shlesinger, M.F. 1982, Proc. Nat. Aca. Sci. USA, 79, 3380

Pomeau, Y., and Manneville, P. 1980, Comm. Math. Phys. 74, 189

Procaccia, I., and Schuster, H. 1983, Phys. Rev. A28, 1210

Sinai, Ia. G. 1982, in Proc. Berlin Conf. on Math. Problems in Theoretical Physics, R.S. Schrader et al. eds., Springer, p. 12.

Voss, R. 1983, Phys. Rev. Lett. 50, 1329

SINGULAR SYSTEM ANALYSIS WITH APPLICATION TO DYNAMICAL SYSTEMS

D S BROOMHEAD, R JONES, G P KING AND E R PIKE

1. INTRODUCTION

Recent work by Bertero et al [1,9] has extended the concepts of the classical information theory of Nyquist and Shannon. They introduce the singular value spectrum to generalise the eigenvalue spectrum of these previous theories. They also quantify the number of degrees of freedom analogous to the Nyquist or Rayleigh resolution limits in more general data analysis problems. The origins of the mathematical theory date back to the turn of the century and the bi−orthogonal series of Picard. The power of normal eigenvalue methods has meant that the application of "singular system" theory to physical problems has been sparse, although Karhunen−Loeve transformations in statistics use closely related techniques, as do recent radar algorithms such as MUSIC. However, a recent application of singular system theory has been the reconstruction of topological phase portraits from one or higher dimensional time series data [3,4]. The natural manifold structure underlying the phase portraits leads to the use of singular system methods in the description of their local geometry [11]. In particular, the dimension of the manifold can be extracted from the data.

In section 2 we review the classical theory of information based upon an eigenvalue inversion in an L^2 space of a first kind Fredholm equation. In section 3 we show how this needs to be generalised for other mapping inversion problems and outline the recent work of Bertero et al, developing a singular system theory to calculate resolution limits or numbers of degrees of freedom for various noninvertible mappings which arise in instrumental science. The effects of finite sampling and truncation of data will be incorporated in section 4 where the theory is described as it applies to vector spaces.

In section 5, we follow Takens [13] in constructing linear maps from the data to lower dimensional spaces using "windowed" sequences of observations. This work is based on a theorem of Whitney [14] which gives the existence of an embedding in a linear vector space of minimum dimension $2m+1$ of the m−dimensional manifold which the phase space trajectory explores.

In the final section we describe how the local dimension of the manifold can be extracted by introducing singular systems for mappings of the data into neighbourhoods of points on this manifold. We then illustrate the technique with some numerical examples.

2. CLASSICAL THEORY OF INFORMATION

The early work of Shannon [10] and others in the field of band−limited communication made it clear that in real physical experiments we must limit the amount of "information" we demand from finite data. Further, they quantified this information in various similar ways as a resolution limit or as a number of degrees of freedom of the experiment (Shannon number). A classic example in which this was already clear was that of the Rayleigh resolution limit in diffraction limited imaging; the Shannon number in this case would be the number of Rayleigh resolution elements covering the object to be imaged.

A detailed analysis of these two problems, which mathematically are one and the same, had to await the work of Slepian and Pollak [12] in the early sixties. These authors solved the problem by expansions in eigenfunctions and introduced the natural basis of prolate spheriodals for an explicit solution to be found for the first time. To the surprise of some the fact that all frequencies (spatial in imaging, temporal in communication) above a certain band limit or cut−off, say Ω, are completely stopped by the apparatus does not deter the mathematician from finding them again in the data and thus reconstructing exactly the full input. This "magic" trick relies on the fact that the reconstruction is one of an entire function and thus partial knowledge can be utilised, by analytic continuation, to find the whole function.

For illustration we confine ourselves to one dimension and pose the problem of band−limited (diffraction−limited) imaging in the form of a Fredholm equation of the first kind. An "object" f(x) gives rise, after transmission through a band−limiting apparatus which passes all Fourier frequencies in full up to a limit Ω and annihilates the rest, to an "image" g(y):

$$g(y) = \frac{1}{2\pi}\int_{-\Omega}^{\Omega} e^{-iky} \left[\int_{-X/2}^{X/2} e^{ikx} f(x)\, dx \right] dk$$

$$= \int_{-X/2}^{X/2} \frac{\sin\Omega(y-x)}{\pi(y-x)} f(x)\, dx \qquad (1)$$

Slepian and Pollak found the eigenfunctions, φ_n, and eigenvalues, λ_n, of this problem, which obey

$$\int_{-X/2}^{X/2} \frac{\sin\Omega(y-x)}{\pi(y-x)} \varphi_n(x)\, dx = \lambda_n \varphi_n(y) \qquad n=0,1,2,\ldots \qquad (2)$$

or

$$K\varphi_n = \lambda_n\varphi_n \qquad (3)$$

by use of a first "miracle" in this field, according to Grunbaum [6], that a commuting differential operator can be found for the integral operator, K, from $L^2(-X/2,X/2)$ into itself defined by

$$g(y) = (Kf)(y) = \int_{-X/2}^{X/2} \frac{\sin\Omega(y-x)}{\pi(y-x)} f(x)\, dx \qquad -X/2 \leqslant y \leqslant X/2 \qquad (4)$$

(other miracles were performed later!).

The formal solution, then, for the reconstruction f(x) is:

$$f(x) = \sum_{n=0}^{\infty} \frac{1}{\lambda_n} \left[\int_{-X/2}^{X/2} g(y)\varphi_n(y)\, dy \right] \varphi_n(x) \tag{5}$$

which, writing the scalar product in L^2 as $<*,*>$, may be expressed in the form

$$f(x) = \sum_{n=0}^{\infty} \frac{\langle g, \varphi_n \rangle}{\lambda_n} \varphi_n(x) \tag{6}$$

The φ_n are a complete orthonormal basis set in $L^2(-X/2,X/2)$ called prolate spheroidal functions. The eigenvalue spectrum λ_n has the well known spectacular behaviour of holding up nearly to unity from $n=0$ until $n = X\Omega/\pi = S$, the "Shannon number", when it drops almost to zero for all higher values of n.

The experimental consequence is that in a decomposition of f into its components φ the data will contain the projection of f in a subspace of $L^2(-X/2,X/2)$ spanned by φ_0 to φ_S without significant distortion or loss, but that the projection of f in the orthogonal complement to this subspace will be severly attenuated on transmission and, in practice, lost in inevitable physical noise. Analytically, of course, in the absence of noise the low amplitude of the higher eigenvalues has no consequence and a full inversion is formally available.

A description in terms of resolution is given by considering the reconstruction $S(x,x_0)$ up to the "lost" components of a δ-function object, say at x_0. Thus

$$\langle g, \varphi_n \rangle = \langle Kf, \varphi_n \rangle = \langle f, K^*\varphi_n \rangle = \lambda_n \langle f, \varphi_n \rangle = \lambda_n \varphi_n(x_0) \tag{7}$$

and hence

$$S(x, x_0) = \sum_{n=0}^{S} \varphi_n(x_0)\varphi_n(x) \tag{8}$$

This "impulse response function" may be computed in any given case and has the form of a central lobe whose width decreases to zero as $S\to\infty$, flanked by smaller side lobes. Some measure of the width of the central lobe can be used as the resolution available. An integral operator with this kernel maps $L^2(-X/2,X/2)$ onto itself in a linear, continuous, computable, consistent and optimal way [5].

3. THE THEORY OF SINGULAR SYSTEMS

In the previous section the problem addressed and solved concerned a linear mapping of an L^2 space into itself. In many cases, however, we are faced with the inversion of mappings from one space, say X, into another, say Y. For example, in the diffraction-limited imaging problem we may wish to record data outside the geometrical image. In this case our integral operator maps $X = L^2(-X/2,X/2)$ into $Y = L^2(-\infty,\infty)$. More generally data is truncated out of necessity at some limits, say [c,d], and the object defined over others, say [a,b], and we thus have a mapping from $L^2(a,b)$ into $L^2(c,d)$. These problems do not have solutions in terms of eigenfunction expansions since an eigenfunction loses its definition except in relation to a mapping of a space into itself.
Fortunately, the required extension of the theory of eigenfunction decomposition has existed since the turn of the century [7] in the form of a bi-orthogonal relationship between bases in the two spaces (the so-called "shifted" eigenvalue problem). Thus

suppose we have a linear bijective mapping K: X→Y, and an adjoint operation K^* such that

$$\langle Kf, g\rangle_Y = \langle f, K^*g\rangle_X \tag{9}$$

where Dirac's notation for the Hermitian conjugate and dual spaces is again used, and the scalar products in X and Y are denoted by $<*,*>_X$ and $<*,*>_Y$, respectively. We now have the following generalisation of equation (3),

$$\begin{aligned} K\,|u_n\rangle_X &= \alpha_n\,|v_n\rangle_Y \\ K^*\,|v_n\rangle_Y &= \alpha_n\,|u_n\rangle_X \end{aligned} \tag{10}$$

where the $|u_n\rangle_X$ and $|v_n\rangle_Y$ are complete orthonormal basis sets in X and Y, respectively, and the α_n are the "singular values" of the decomposition.

By substituting the second equation of (10) into the first, and vice versa, we have

$$\begin{aligned} KK^*\,|v_n\rangle_Y &= \alpha_n{}^2\,|v_n\rangle_Y\,, \\ \text{and} \quad K^*K\,|u_n\rangle_X &= \alpha_n{}^2\,|u_n\rangle_X\,. \end{aligned} \tag{11}$$

Thus the $\alpha^2{}_n$'s are the eigenvalues of the operator KK^*:Y→X and also of the operator K^*K:X→Y, with $|v_n\rangle_Y$ and $|u_n\rangle_X$ the respective eigenfunctions.

We call the triple of mapping, object space, and image space $\{K,X,Y\}$ a singular system, with natural singular value spectrum and singular bases $\{\alpha_n;u_n,v_n\}_{n=0\ldots\infty}$.

The use of the singular system to invert a Fredholm equation of the first kind,

$$g(y) = \int_a^b K(x,y)\, f(x)\, dx \qquad c \leqslant y \leqslant d \tag{12}$$

is a simple generalisation of the solution (6) for the "square" problem of the previous section. We have thus

$$f(x) = \sum_{n=0}^{S} \frac{\langle g, v_n\rangle_Y}{\alpha_n}\, u_n(x) \tag{13}$$

where the series expansion in X in the basis functions u with coefficients given by scalar products in Y with the basis functions v is truncated (or may be rolled off more smoothly if desired) at a value of n which is a generalisation of the Shannon number. This new number of degrees of freedom is not defined quite so sharply as is possible with the eigenvalue spectrum of the prolate spheroidal functions, where the level of noise present hardly affected the issue. In fact, the concept of number of bits of information and the idea of an information theory as such rested on the remarkable properties of this spectrum as discussed above and hence was formulated without explicit reference to noise levels. We now have to decide at what point the transmitted singular component will be lost in noise and this will depend on how the singular value spectrum falls off (which it must if K is compact) and on the actual noise levels present.

The most important property of this method of data reduction lies in the fact that the information in the coefficients of the singular functions is optimally compressed. That is to say that if a series of a fixed number of components is to be used to approximate the function, the least error will ensue if the singular basis is used. In the pattern recognition field the "representational entropy" is said to be a minimum, or in rougher terms, the "moment of inertia" of the spectrum about the origin is the least possible.

4. SAMPLED AND TRUNCATED DATA

In real life we are not able to use continuous functions in an L^2 space to describe our data exactly. The nature of physical experimentation is such that data will be sampled with a finite resolution and truncated over some reasonable limits. In other words the output of an experiment will be a vector in an N−dimensional vector space rather than a L^2 function.

Again, fortunately, the singular system theory can be adapted so that K is a mapping from an L^2 space X into a vector space Y. On the other hand, if the input happens to be best described as a vector in a vector space X, then K may be a rectangular matrix mapping $X \to Y$. In the first case K^*K is a finite rank integral operator and KK^* a square matrix. In the second case both KK^* and K^*K are square matrices (of different dimensions).

The same inversion procedure is applied and equation (13) again gives the result with $<*,*>_Y$ defined appropriately as a Euclidean scalar product and with the u_n either L^2 functions (represented in actual computer calculations on an arbitrarily fine mesh) in the former case or suitable vectors in the latter case.

Many inversion problems have been treated by these methods in recent years [2,8] and the technique gives both an excellent match to digital experimental data and an exact rejection of out−of−band noise. Furthermore, the technique lends a comprehension of the nature of the ill−conditioning which would arise if the generalised Shannon number were taken at too high a value. We might hope that in a well planned experiment we are not asked to recover "invisible" singular components which have been transmitted so weakly (if the value of α_n is so small) that the information required is masked by noise in the data. If this is not the case, then a reconstruction can only be given of that part of the object which lies in the "visible" projection on the subspace spanned by u_0 to u_S where S is the generalised Shannon number.

5. TIME SERIES DATA

The application of the above theory to dynamical systems rests on the fact that a manifold M of dimension m, on which the phase space trajectory evolves, can be embedded in a linear vector space, Y, of sufficiently large dimension. An embedding is a diffeomorphism between the manifold M and an image manifold in Y. The foundation of this is a theorem by Whitney [14] who proved that a smooth (C^2) m−dimensional manifold which is compact and Hausdorff may be embedded in R^{2m+1}. The dimension 2m+1 is, generically, the minimum required to ensure the embedding. Its origin can be understood without going into the detail of the proof: It can be shown that there exists an n such that M can be embedded in R^n. The proof proceeds by looking at the set of possible projections from R^n to R^{n-1} to see if they give an embedding in R^{n-1}. A projection can be characterised by the unit vector along which it acts. Therefore the set of all projections $R^n \to R^{n-1}$ can be associated with the unit sphere S_{n-1} in R^n. Of these projections there will be a set which do not give an embedding because they superimpose originally distinct points of M. These projections are characterised by the unit vectors $(x-y)/|x-y|$, where x and y are distinct points of M. Thus the image of the map $\Phi: M \times M - \Delta \to S_{n-1}$, where

$\Phi(x,y)=(x-y)/|x-y|$ and Δ is the diagonal of $M\times M$, is the subset of S_{n-1} which corresponds to projections that do not give an embedding. The dimension of $M\times M-\Delta$ is 2m, and that of S_{n-1} is $n-1$, thus if $2m<n-1$ the image of Φ is nowhere dense in the unit sphere. In other words if $n>2m+1$ "most" projections can give an embedding. Conversely, in $\mathbb{R}^{2m+1}$ "most" projections will not give an embedding in $\mathbb{R}^{2m}$. Note that the proof makes no assumptions about the detailed structure of M, thus some m−manifolds can be projected into fewer than $2m+1$ dimensions.

As an example consider a closed 1−manifold (a limit cycle, for instance). The theorem states that it needs to be embedded in $\mathbb{R}^3$, but most projections of a circle in $\mathbb{R}^3$ will give an embedding in a plane. However, general 1−manifolds in $\mathbb{R}^3$ will be knotted. For these there is no projection that does not give a self−intersecting image in the plane.

In previous work Broomhead and King [3,4] have used the singular system analysis to construct an embedding of the global attractor. The approach in that work considers time series data x_i of $(N+n)$ samples at equal time steps and constructed from it N−dimensional vectors in $\mathbb{R}^N$ (which is to be associated with the space X of the previous sections) with basis set e_i given by

$$
\begin{aligned}
|e_1\rangle &= (1,0,0,\ldots,0)^T \\
|e_2\rangle &= (0,1,0,\ldots,0)^T \\
&\;\;\vdots \\
|e_N\rangle &= (0,0,\ldots,0,1)^T
\end{aligned}
\tag{14}
$$

Note that a point in $\mathbb{R}^N$

$$|g_i\rangle = (x_i, x_{i+1}, \ldots, x_{i+N-1})^T \tag{15}$$

effectively represents the entire time series.

Takens proved that a set of n consecutive values from the time series give a vector on the image manifold in $\mathbb{R}^n$ (which is to be associated with the space Y of the previous sections) provided that $n\geqslant 2m+1$. Thus the ith point of the trajectory in $\mathbb{R}^n$ is denoted by

$$|y_i\rangle = (x_i, x_{i+1}, \ldots, x_{i+n-1})^T \;. \tag{16}$$

To describe the evolution of the trajectory we write

$$\langle y_i| = \langle e_i|K \tag{17}$$

where K is an $N\times n$ matrix

$$K = \begin{bmatrix} y_1^T \\ y_2^T \\ \cdot \\ \cdot \\ \cdot \\ y_N^T \end{bmatrix} \tag{18}$$

This "trajectory" matrix has the further properties that

$$\langle G| = \langle g_i|K \quad , \tag{19}$$

where the kth component of G is the autocorrelation coefficient $\Sigma x_i x_{i+k-1}$ and

$$K|f_i\rangle = |g_i\rangle \quad , \tag{20}$$

where the $|f_i\rangle$ are the unit basis vectors in $\mathbb{R}^n$. Equation (20) is of the form discussed in the previous sections and we may thus use the singular system theory to find the "information" content of the data by projecting it onto the singular vectors of the operator K. In this case the smaller n×n matrix K^*K is obviously the best to diagonalise and is, as can quickly be verified using eqn (19), the matrix of correlation coefficients averaged over n consecutive samples of the time series. This matrix is the Karhunen–Loeve transformation which is used in statistics to obtain statistically independent processes.

As before, the number of singular values, α_n, above a given "noise floor" will tell us the number of degrees of freedom of the problem, and each singular component will be orthogonal and statistically independent from the others. However, the number of degrees of freedom defined in this way is not an invariant of the embedding process. A suitable invariant is the dimension of the manifold. In the next section we will apply singular system methods to calculate this number.

6. THE DIMENSION OF THE MANIFOLD

Topologically irrelevant deformations of the manifold can change the apparent number of degrees of freedom. Locally, however, an m–dimensional manifold can always be approximated by the linear vector space $\mathbb{R}^m$. Thus we may carry the analysis further to obtain the local dimension of the manifold by mapping not the whole trajectory from $\mathbb{R}^N$ to $\mathbb{R}^n$ but only those parts of it which lie in a sufficiently small neighbourhood of some point chosen on the manifold [11].

The idea is firstly to reduce the problem by projecting the trajectory onto the "deterministic" subspace by shedding the noise subspace in the usual way. Secondly, to select a point on this trajectory and to construct a neighbourhood matrix K' by rejecting those vectors which lie outside a ball of given radius r centred at this point (using the Euclidean norm). The singular values of K' are now widely separated into a set corresponding to the topological dimension of the manifold and a set corresponding to the residual noise. The singular values in the first set scale linearly with r and hence may be identified easily. Indeed, a simple calculation shows that for a manifold of m dimensions the m singular values are all equal and given by

$$\alpha = \left[\frac{N_b}{m+2}\right]^{\frac{1}{2}} r \quad , \tag{21}$$

where N_b is the number of points used in the neighbourhood.

We illustrate the above method with three examples: a 1–torus, 2–torus, and a 3–torus. We have generated incommensurate sinusoids and passed them through a cubic nonlinearity of the form $x + x^3$. The data were sampled at 100 times the largest fundamental frequency. The resultant time series (quantized to 2^{14} levels) are shown in Fig 1. Classical discrete Fourier transformation (with Hanning weights) of each of these time series produce the complicated spectra shown in Fig 2. In each of these spectra the quantization noise floor can be seen.

The time series were windowed to produce sequences of vectors in $\mathbb{R}^{25}$. The corresponding singular spectra are shown in Figs 3–5, where there are respectively four, five, and five singular values above a band–limited noise floor; the noise being

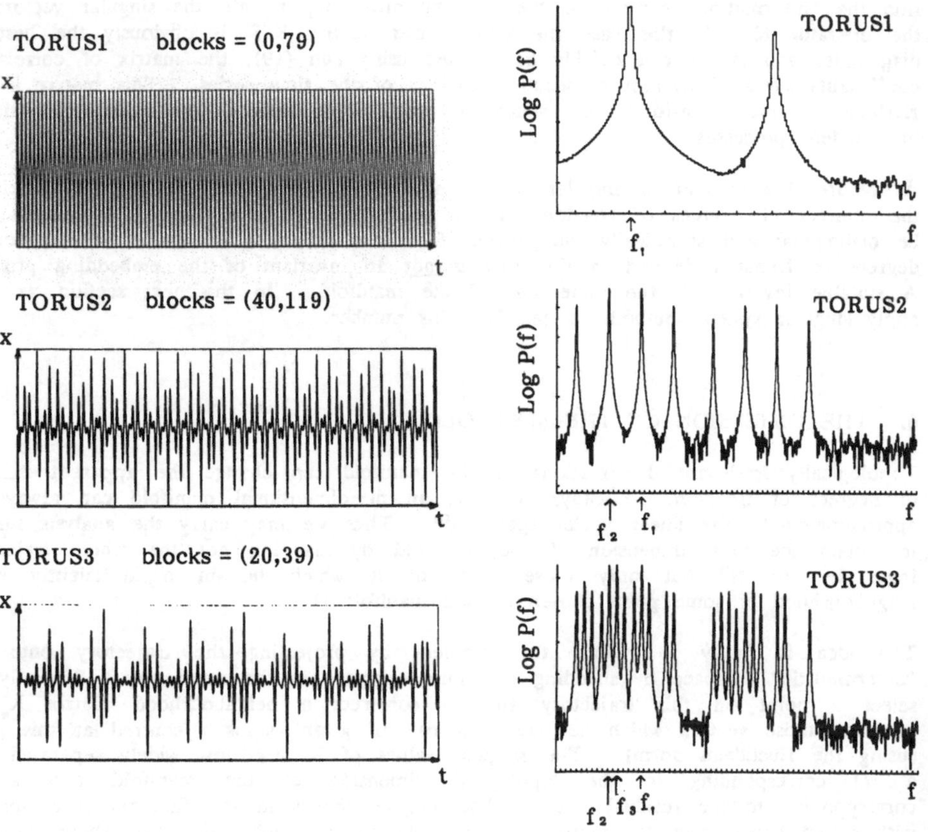

Figure 1. Time series generated by passing sinusoidal signals through a cubic non-linearity.

Figure 2. Power spectra constructed from the time series shown in fig. 1. The frequencies marked in the figure are $f_1 = 1.0$, $f_2 = (\sqrt{5} - 1)/2$, and $f_3 = 1/\sqrt{2}$.

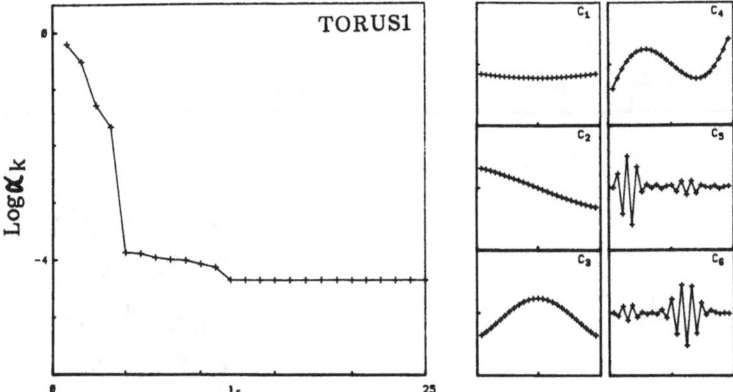

Figure 3. Global singular value spectrum and their associated singular vectors for the 1−Torus. The six singular vectors correspond to the six largest singular values.

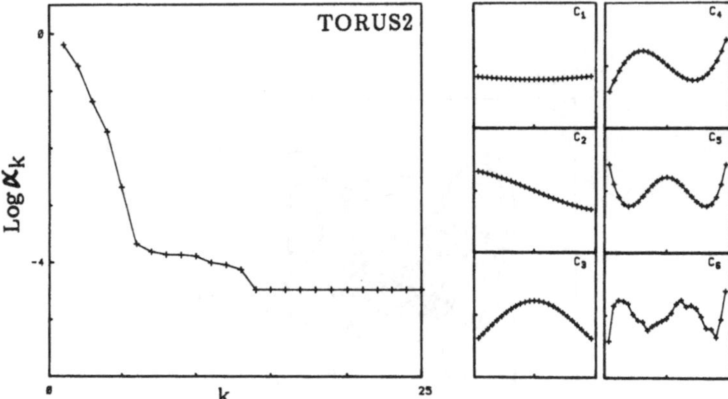

Figure 4. As Fig 3 but for the 2−Torus.

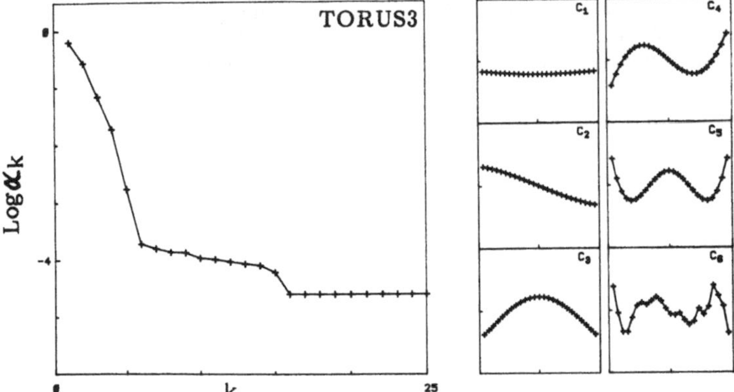

Figure 5. As for Figs 3 and 4 but for the 3−Torus.

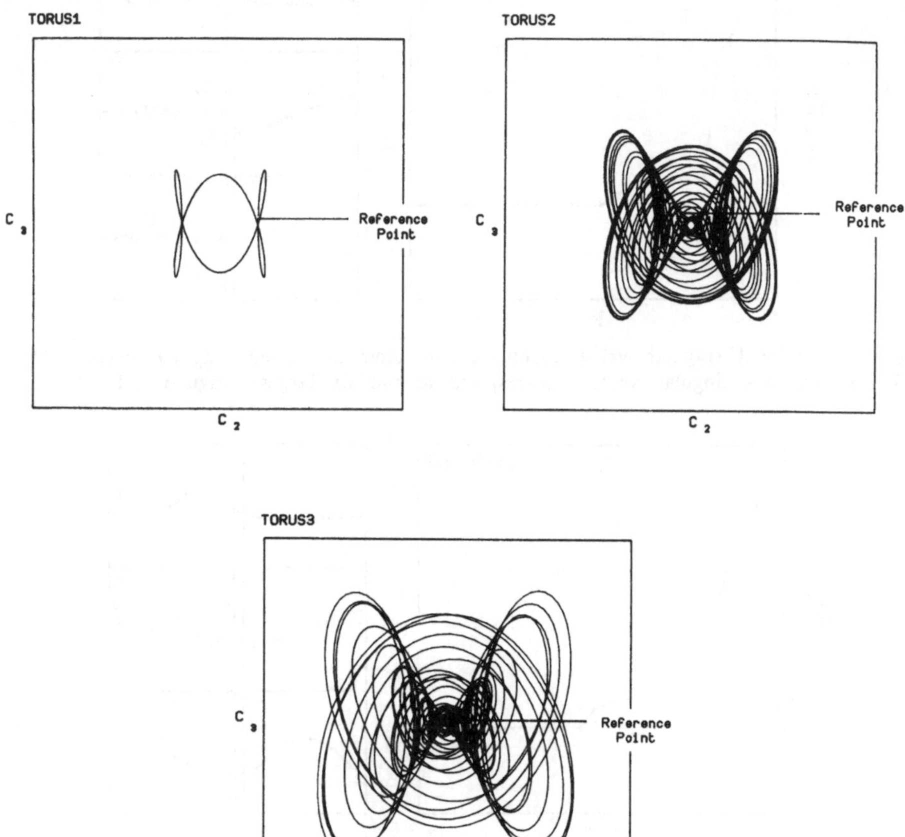

Figure 6. Phase portraits reconstructed from the time series data shown in Fig 1. The points marked in the above figures were used as reference points in the calculation of the topological dimension of the Tori (see Figs 7−9).

dominated by the quantisation of the data. Also shown in each figure are the singular vectors associated with the six largest singular values. Phase portraits constructed by projecting K onto the second and third singular vectors (c_2 and c_3) are shown in Fig 6.

In each case when performing the local analysis, we assume the deterministic subspace to be spanned by the first five singular vectors. The reference points used in the calculations are those shown in Fig 6. The results of this analysis are summarised in Figs 7–9. These figures show a log–log plot of the singular spectrum of the neighbourhood matrix [illegible] as a function of the radius of the ball defining the neighbourhood. In Fig 7 we see a singular value which scales linearly with radius, and which is well separated from the rest which do not. A single singular value scaling linearly with r implies that m=1 for this data. This is the expected result. If the neighbourhood can be chosen small enough, then locally the limit cycle would appear linear and would be well approximated by its tangent at the point. Thus the singular vector corresponding to the isolated singular value will be colinear to the tangent. As the neighbourhood is allowed to increase in size, then providing the local curvature does not vanish we expect a second singular value scaling as r^2 to appear. (The singular vector associated with this will be normal to the curve.) This can be seen in Fig 7.

The results for the 2–torus shown in Fig 8 have a similar interpretation. In this case there are two singular values that scale linearly with r and thus imply the expected m=2. Since the singular value that scales as r^2 is apparent at the smallest r we were able to probe with the length of time series used (approx 10^6 samples), we were never free from the effects of curvature of the manifold. It is to be expected that this situation will worsen as the dimension of the manifold increases. This is illustrated by the results for the 3–torus shown in Fig 9. Although we can see three singular values scaling linearly with r, they are no longer well separated from the rest.

It is interesting to contrast the results of this local analysis with the Fourier spectra shown in Fig 2. Due to the nonlinearity the latter show many harmonics which makes it difficult to pick out the number of fundamental frequencies contributing to the specta. On the other hand the dimension of the manifold determined from the local analysis remains at one, two and three, respectively (i.e., the number of original incommensurate sinusoids).

CONCLUSIONS

We have reviewed the theory of singular system analysis, a generalisation of eigenfunction analysis, with particular reference to its use for extending the concepts of classical information theory. The measures of information content or number of degrees of freedom which the theory provides have been used to analyse time series data from nonlinear dynamical systems. The number of degrees of freedom that results from the singular system analysis identifies the dimension of a deterministic subspace. A local analysis in this reduced space was then used to discover the underlying dimension or number of independent processes contributing to the motion. It was found that an excellent measure of the topological dimension of the manifold in which the system evolves can be determined.

We have seen that singular system analysis offers a new direction in the treatment of data from that of conventional Fourier or correlation processing. The full potential of these methods have yet to be explored, but we feel it holds great promise.

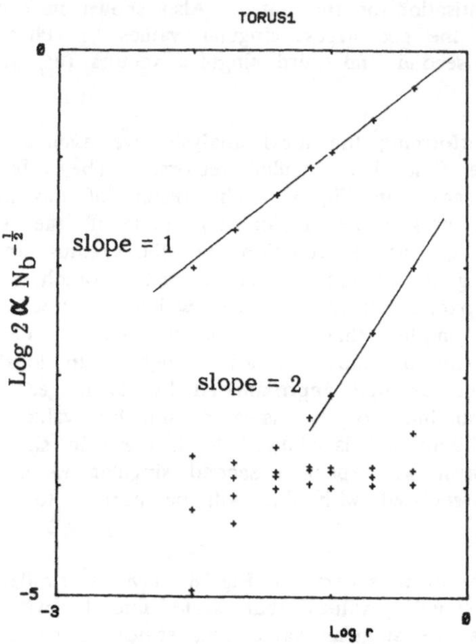

Figure 7. The local singular value spectrum as a function of the neighbourhood radius for the 1−Torus. Only one singular value scales linearly with r implying m=1.

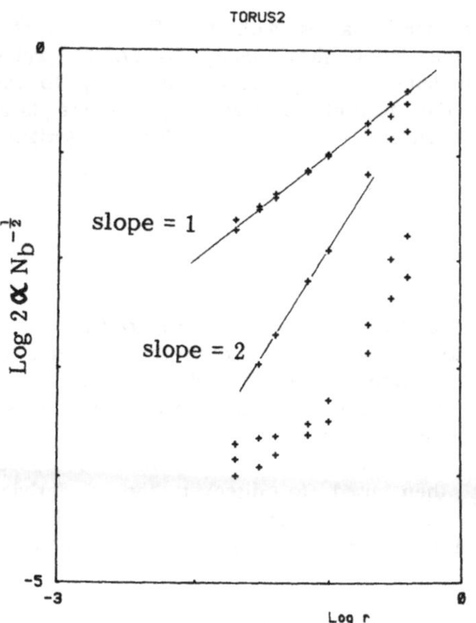

Figure 8. The local singular value spectrum for the 2−Tori showing two singular values scaling linearly with r.

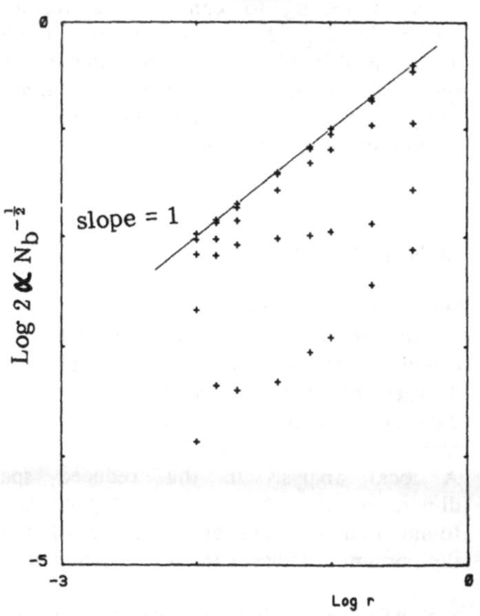

Figure 9. The local singular value spectrum for the 3−Tori showing three singular values scaling linearly with r.

REFERENCES

[1] M Bertero, C De Mol and E R Pike (1985), Inverse Problems, 1, 301.

[2] M Bertero, P Brianzi, E R Pike, G de Villiers, K H Lan and N Ostrowsky (1985), J Chem Phys, 82, 1551.

[3] D S Broomhead and G P King (1986), Physica 20D, 217.

[4] D S Broomhead and G P King (1986), *Nonlinear Phenomena and Chaos*, ed S Sarkar (Adam Hilger, Bristol) p113.

[5] C L Byrne and R M Fitzgerald (1982), SIAM J Appl Math, 42, 933.

[6] A Grunbaum (1986), "Inverse Problems" Fondazione CIME Procs Montecatini (Springer–Verlag).

[7] E Picard (1910), R C Mat Palermo, 29, 615.

[8] E R Pike, J G McWhirter, M Bertero and C De Mol (1984), Proc IEEE, 131, 660.

[9] E R Pike (1986) "Inverse Problems" Fondazione CIME Procs Montecatini (Springer–Verlag).

[10] C F Shannon and W Weaver (1949), *The Mathematical Theory of Communications*, (Univ of Illinois Press, Chicago).

[11] D S Broomhead, R Jones and G P King (1986) submitted.

[12] D Slepian and N O Pollak (1961), Bell System Technical Journal, 40, 43.

[13] F Takens (1981), Detecting strange attractors in turbulence, Lecture Notes in Mathematics, eds D A Rand and L–S Young (Springer, Berlin) p366.

[14] H Whitney (1936), Ann Math, 37, 645.

A REVIEW OF PROGRESS IN THE KICKED ROTATOR PROBLEM

G CASATI

ABSTRACT

The classical motion of a rotator under an external time-periodic, δ-like perturbation has been studied in great detail in the last two decades. Indeed this system, as the external perturbation is increased, displays the great variety and complexity of generic, non integrable, classical systems. The simplicity and richness of this model makes it a good candidate to investigate the manifestation of classical chaotic motion in quantum mechanics. In this paper we discuss the quantum behaviour of such system. The analytical and numerical results so far obtained indicate that quantum mechanics places strong limitations to classical chaotic motion. This is the main feature of quantum motion and, even if some statistical behaviour is present, a satisfactory classification of its statistical properties, as we have in classical mechanics, is here lacking.

Moreover, as for the classical case, the qualitative properties of the quantum motion appear to depend also on some fine details such as the number theoretic properties of the ratio between the frequency of the unperturbed motion and the frequency of the external perturbation.

We also discuss the relation between the quantum limitation of classical chaos and the Anderson localization in static potentials which is due to quantum interference effects.

1. INTRODUCTION

Our understanding of the qualitative behaviour of classical dynamical systems has recently been improved by the discovery of the so-called chaotic motion. For centuries, after the discovery of Newton's equations of motion, it has been substantially impossible to go beyond the two-body Kepler problem and systems of harmonic oscillators. Pertubation theory has been the main tool used to obtain information on the behaviour of many body systems interacting via non linear forces, but it was soon realized that resonance phenomena rendered the traditional perturbation methods divergent. In this connection great significance was attached to the Poincare proof that, under very general conditions, any perturbation of an integrable system will destroy all the analytical constants of the motion except the energy. This was considered as a convincing argument in favour of the ergodic hypothesis and to the justification for equilibrium classical statistical mechanics. In recent years, two major achievements have been made which led to a new and better understanding of the qualitative features of classical motion:

i) the mathematical papers of Kolmogorov, Arnold and Moser [1] (KAM theorem) and the pioneering analytical and numerical works of Chirikov [2], Ford [3], Henon [4] have shown that, contrary to the previous general belief, small perturbations of an integrable system leave the system close to the integrable system and most orbits remain quasi-periodic. However, as the strength of the perturbation is increased, the system undergoes a transition from near-integrable to ergodic motion. In this latter case the orbits wander freely on the energy surface and statistical methods can be applied;

ii) classical dynamical systems, governed by purely deterministic laws, may exhibits a purely random motion despite the seeming contradiction of these terms [5]. These systems are characterized by exponential divergence of initially close orbits, namely by positive maximal Liapounov exponent λ

$$\lambda = \lim_{t\to\infty} 1/t \ln (d(t)/d(o))$$

where d(t) is the distance between two initially close orbits or, more precisely, the modulus of the linearized solution of the equations of motion. Algorithmic complexity theory [5] shows that if $\lambda>0$ then almost all orbits are random, unpredictable and uncomputable. This fact may be understood in terms of symbolic dynamics. The sequence of symbols or number X_n which specifies the trajectory (n is integer time) has positive algorithmic complexity which means that the length of the algorithm necessary to reproduce the sequence X_n increases like n. In other words, in order to specify the orbit, one must know it in advance. In this sense the motion is truly random.

It must be remarked that this type of motion, exponentially unstable with respect to initial conditions, is structurally stable which leads to stability of averaged quantities. We stress also the fact that randomness in the motion does not necessarily imply exponential decay of correlations; on the contrary, it can be shown [6] that decay of correlations have a long-time tail with power-law decay.

Between the two extreme types of motion described above, integrable or near-integrable and completely random, there is a hierarchy of statistical properties (ergodicity, weakly-mixing, mixing). These properties are distinguished by the nature of the spectrum of the Liouville operator on the energy surface. In particular, ergodic systems may have discrete spectrum and therefore, while they justify equilibrium statistical mechanics, they do not exhibit an approach to statistical equilibrium. For this purpose one needs a continuous spectrum which characterizes mixing systems.

The above qualitative considerations show that classical systems have a rich variety of different behaviour. A natural question now arises whether and to what extent this richness of classical motion survives in quantum mechanics.

At first sight it appears that there is nothing in the solution of the Schroedinger equation so complex as the orbit of a classical chaotic system. Indeed a necessary (not sufficient) condition for chaotic motion in classical systems is continuous spectrum of the motion while, if one consider bounded, conservative, finite particle number systems, the energy spectrum is discrete no matter whether the corresponding classical systems are chaotic or not and this implies almost-periodicity in time of the wave-function and therefore absence of chaotic and irreversible behaviour in such systems. On the other hand, on the basis of Ehrenfest's theorem, one expects that, in the semiclassical region, a narrow packet will follow the classical trajectory and one is led to suspect that, at least, quantum systems have different qualitative properties depending on whether the corresponding classical systems are integrable or chaotic. The study of possible manifestations of chaos in such systems was initiated by I.C. Percival [7]. He proposed that in the semi-classical limit, the quantal energy spectrum should consists of two parts with strongly contrasting properties: a regular and an irregular part. He also remarked that "the distribution of levels of the irregular spectrum could take on the appearence of random distribution". Later the statistical distribution of levels has been extensively studied and it has been found that the distribution of eigenvalues of classically chaotic systems fall in different universality classes depending on whether the dynamics possesses a time-reversal invariance or not. These distributions have the same statistical properties of the eigenvalues of random matrices (GOE, GUE) and do not depend on the dynamics of the system considered nor on the type of interaction (nuclear, electromagnetic ecc.). However for an update presentation of this subject see ref. 8.

In order to search for some kind of randomness in the time evolution of a quantum system, one must consider systems acted upon by external time-periodic perturbations. Indeed for such systems the solution of Schroedinger Eq. may be written in the form

$$\psi(t) = P(t)\, e^{iGt}\, \psi(o)$$

where P is periodic and G is self-adjoint. Here the spectrum of the evolution operator over one period may be continuous and therefore there is the possibility, in principle, of some chaotic motion.
Following this line of thought we considered a time-periodic perturbed system which is sufficiently simple but which display the typical extremely rich behaviour of classical systems: the δ-kicked rotator. In classical mechanics, the study of the δ-kicked rotator has provided deep insight into the general behaviour of dynamical systems since it shares almost all their main features. Correspondingly, we believe that the study of the quantum properties of the kicked rotator, expecially in regions of parameters when the corresponding classical model is chaotic, will be of great significance for understanding of the qualitative features of the quantum motion.

2. THE CLASSICAL ANALYSIS

Let us consider the classical Hamiltonian

$$H = p^2/2 + \omega^2 \cos\theta \sum_j \delta(t-jT), \tag{1}$$

where p is the rotator momentum, θ is the angular coordinate, T the kick period and ω the perturbation strength.

By expanding the δ-function in Fourier series we have

$$H = p^2/2 + (\omega^2/T) \cos \theta \, (1+2\sum_{n=1}^{\infty} \cos (2\pi nt/T). \tag{2}$$

The perturbation strongly affects the motion when the phase is slowly varying, namely at resonance values

$$\dot{\theta} = p_n = 2\pi n/T \tag{3}$$

Following Chirikov's analysys [2], we describe the motion in the vicinity of each resonance by expanding the unperturbed Hamiltonian near p_n up to second order terms in p and by introducing new coordinates I, φ using the time dependent generating function

$$F(I, \theta, t) = (p_n + I)(\theta - 2\pi nt/T). \tag{4}$$

In terms of new variables $I = p - p_n$, $\varphi = \theta - p_n T$ one may show that the Hamiltonian near p_n becomes the pendulum Hamiltonian

$$H = I^2/2 + (\omega^2/T) \cos \varphi. \tag{5}$$

The ampitlitude of phase oscillations $(\Delta I)_n$, called the resonance half-width, is given by half of the maximum distance between the pendulum separatrices [2]

$$(\Delta I)_n = 2\,(\omega^2/T)^{1/2} \tag{6}$$

On the other hand, the distance between two consecutive resonances is

$$\delta_n = I_{n+1} - I_n = 2\pi/T \tag{7}$$

If the perturbation ω^2/T in the Hamiltonian (5) is sufficiently small, the distance δ_n between resonances is larger than the width $(\Delta I)_n$ and the motion is trapped inside the resonance island. As the strength of the perturbation increases, the resonance width grows larger and larger until a critical value is reached, at which two consecutive resonances overlap. The critical value is determined by the relation $(\Delta I)_n = \delta_n/2$ which gives

$$\omega^2 T = \pi^2/4. \tag{8}$$

When the perturbation exceedes the critical value given by (8), the resonance islands overlap and the orbits move freely in action space throughout the island chain region in a random like manner. In this situation the rotator moves as under the action of some random force even if the latter is, in fact, completely deterministic and periodic in time.

The estimate (8) gives the correct critical value in order of magnitude, but not the exact value, since we neglected secondary effects like second order resonances etc. [2]. Numerical computations confirm the above estimate roughly within a factor two. Indeed, due to the presence of the δ-function, the classical equations of motion can be integrated and reduced to the mapping.

$$\begin{aligned} P_{n+1} &= P_n + K\sin\theta_n, \\ \theta_{n+1} &= \theta_n + P_{n+1}, \end{aligned} \tag{9}$$

where $K = \omega^2 T$, n is time measured in number of kicks and P is the dimensionless angular momentum $P_n = p_n T$.

Mapping (9) is the well-known "standard map", extensively discussed in the literature and frequently used, at a tutorial level, to illustrate the great complexity of motion for simple dynamical systems. For K=0 this mapping is integrable and all orbits lie on smooth curves. For $0<K<K_c$, with $K_c \sim 1$ the previous analysis indicate that most orbits continue to lie on smooth curves and, in particular, the kinetic energy remains bounded with a variation $(\Delta P) \sim K$. On the other hand, when K exceeds the critical value K_c, most of the invariant curves disappear and the mapping orbits become chaotic: the system performs a random walk in momentum space leading to a diffusive growth of the kinetic energy.

$$\overline{P^2} \simeq (K^2/2)n, \tag{10}$$

and to the angular momentum distribution of the Gaussian type

$$f(P,n) = (K^2)\pi n)^{-1/2} \exp(-P^2/k^2 n). \tag{11}$$

A characteristic feature of this motion is the exponential instability of orbits with respect to initial conditions, which leads to a rapid randomization of the phase variable and hence to the diffusive motion described by eqs. (10) and (11).
Without entering in further details, we may certainly state that the motion of systems (1) presents completely different qualitative features depending on whether K is less or larger than $K_c \sim 1$.

3. THE QUANTUM ROTATOR

Let us now turn to the quantum description. Following ref. [9] we consider the quantum Hamiltonian

$$H = - \hbar^2 \partial^2/\partial\theta^2 + \omega^2 \cos\theta \sum_j \delta(t-jT). \qquad (12)$$

By letting

$$\psi(\theta,t) = \sum_n c_n e^{in\theta}, \qquad (13)$$

we may write the solution of the Schroedinger equation as a mapping $\psi \to \overline{\psi}$. For the wave function after one step, which includes a free rotation of time T and a kick, we find

$$\overline{\psi}(\theta) = S\,\psi(\theta) = \exp(-ik\cos\theta)\Sigma_n\, c_n \exp[i(n\theta-2\pi n^2\tau)] \qquad (14)$$

where

$$k = \omega^2/\hbar\,, \quad \tau = \hbar T/4\pi. \qquad (15)$$

From eq. (14) we see that it suffices to consider values of τ within the interval [0,1].
We may now perform numerical iterations of the quantum mapping (14), and

inquire about the properties of quantum motion by varying the classical parameter K below or above $K = K_C$. Notice also that $K=4\pi k\tau$ and that the classical limit is reached by letting $k \to \infty$ $\tau \to 0$ keeping the classical parameter K fixed.

We numerically iterated the quantum mapping starting from a given set of $\{c_n(0)\}$ and computed the probability distribution

$$\rho(n) = |c_n|^2$$

and the average energy

$$<E> = (1/2)\, \Sigma_n\, n^2\, \rho(n)$$

In order to investigate the extent to which the numerically computed quantum distribution $\rho(n)$ mimics the classical stochastic distribution we write the quantum version as

$$f(n) = (k^2 \pi t)^{-1/2} \exp(-\, n^2/k^2\, t)$$

where t is integer time measured in number of kicks periods.

Our first expectation was that for $K>K_c$, in analogy with the classical results, the distribution (11) would be obtained also in the quantum case. With our great surprise this expectation was not verified. Instead, as it is shown in Fig. 1 the quantum system exhibited for typical values of the period τ, a strong stability character.

It was also found that the quantum motion can mimic the classical diffusive energy growth yielding

$$<E> = \sum_m (m^2/2)\, |c_m{}^2 \simeq (k^2/4)t.$$

However, in the quantum motion this linear energy growth persists only up to a *break time* t_b. Empirically it appears that $t_b \to \infty$ as $\hbar \to 0$ (or equivalently $k \to \infty$), for constant $K=4\pi k\tau$. For times greater than the break time, the quantum energy appears to enter a steady-state oscillatory regime. In short, the kicked quantum rotator apparently introduces a limitation to classical diffusion and randomness.

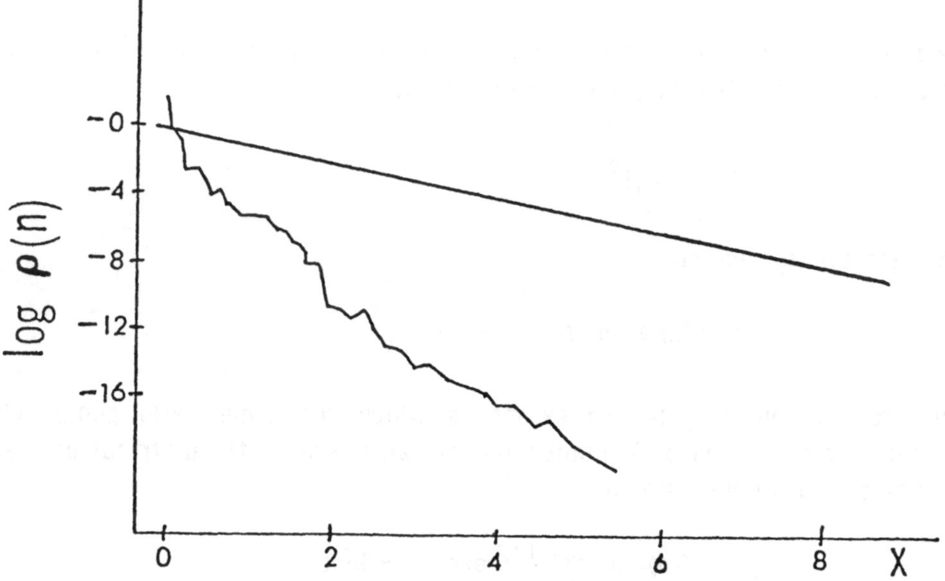

Fig. 1 - A plot of the logarithm of the quantum probability distribution $\rho(n)$ versus the normalized variable $x = n^2/k^2t$ at time $t=25$ for $k = 10$ and $\tau=1/2$. The straight line is a plot of $\ln(e^{-x})$. Here the two curves are quite distinct indicating a lack of stochasticity in the quantum motion even though $K = 5$.

We recall that the character of the time evolution for ψ is known once the quasi energy (abbreviated q.e., hereafter) spectrum is known. It can be shown [9] that, the q.e. spectrum for our model can be pure point only when the rotator energy remains bounded for all times, in which case the initial wave packet remains localized in momentum space. Thus the very existence of a break time after which energy growth stops completely would ensure that the quasi energy spectrum is pure point and that the full quantum motion is almost periodic.

Chirikov, Izrailev, and Shepelyansky [11] have advanced an heuristic argument which yields a quantitative estimate for this quantum limitation of diffusion due to the break-time phenomenon. In essence, this argument is the following: assume that the q.e. spectrum is pure point since otherwise there will be no finite break time. By the quantum theory of measurement, this pure-point character of the spectrum can become apparent only after a certain time t*. Prior to this time the system will behave as if its spectrum were continuous and hence the limitation on diffusion will occur no later than t*. Chirikov et al. therefore identify t* with the break time t_b in order of magnitude.

To estimate t*, recall that the total number of q.e. levels is infinite and that all these levels are located within a bounded interval. In addition, recall that each particular state ψ of the system can be obtained as a superposition of quasi energy eigenstates; while, generally speaking, the superposition involves an infinite number of q.e. states, we may assume that for an initially localized packet, the finite time evolution of ψ is dominated by a finite number N_ψ of them. This effective number N_ψ is also the number of eigenfrequencies actually occuring in the time evolution of ψ. Therefore the spectrum of the motion corresponding to state ψ will consist of N frequencies with average spacings about N_ψ^{-1}. It is now clear that t* $\approx N_\psi$ and we must estimate N_ψ. First, we assume that N_ψ roughly coincides with the number of unperturbed eigenstates significantly involved in the motion. Next, we identify N_ψ with $\Delta P(t^*)$, that is, with the spread in momentum achieved at time t*. Experimentally we have $\Delta P(t^*) \sim k(t^*)^{1/2}$ (for k>1, K>1). Since $N_\psi \sim t^* \sim \Delta P(t^*)$, we finally get that, in order of magnitude, $t^* \sim k^2$. In other words, both the break time and the localization length N_ψ are of the order of the classical diffusion coefficient.

Localization of quantum wave packets in cases where classical mechanics would predict a diffusive behaviour is a well-known phenomenon in solid-state physics also. Indeed, the (time - independent) Hamiltonian of a particle in a random potential on a line can be proved to have a pure-point spectrum with eigenfunctions exponentially localized in space; this fact is known as Anderson localization. Along this line, Grempel, Fishman, Prange [12] gave an important contribution in understanding the quantum kicked rotator problem. They were able to show that the quantum kicked rotator is related to Anderson's problem of motion of a quantum particle in a one-dimensional lattice in a static potential. In this second problem, if the

potential is random, quantum interference effects suppress diffusion and lead to a localization of the wave packet. By establishing a mapping between the two systems they conclude that a similar exponential localization should take place in the quantum rotator also.
More precisely the kicked rotator problem (12) can be mapped into a tight-binding model

$$T_m u_m + \Sigma W_r u_{m+r} = Eu_m \tag{16}$$

where the hopping coefficient W_r is the Fourier transform of $W(\theta)=$ tg $\left((k/2)\cos\theta\right)$, $E = - W_0$ and where the diagonal potential $T_m = \text{tg}\,(\lambda-\tau m^2/2)$. The angular momentum in the quantum problem corresponds to the lattices sites m in the solid state problem. In the tight binding problem, eq. (16) is an eigenvalue equation for the energy E while the quasi-energy λ plays the role of a parameter in the potential.
Whether the similarity with the kicked rotator problem is close enough that Anderson's result can be invoked in this case is a point which turns out to depend on the "degree of randomness" of the number sequence $T_m=\text{tg}(\lambda-\tau m^2/2)$. This is a very delicate question; at first sight one would say that the answer might depend strongly on the arithmetic properties of τ. If T_m is periodic in m (τ rational) then the corresponding eigenstates are Bloch extended states. If T_m is a random sequence then the spectrum is discrete and the states are exponentially localized. In our case, for irrational τ, T_m is not strictly random but can be considered as pseudorandom in the sense that, to some extent, it mimics a random sequence.

As a matter of fact, for typical irrational values of τ, the picture of Anderson localization in momentum space fits very well the rotator problem also. In Fig. 2, for two different close values of τ (expressed in continued fraction), we show the shape of the squared modulus of the wave function in momentum space, after the break time t* and for an initial δ-like state. The exponential behaviour is apparent here; also, the localization length is in excellent agreement with the estimate given above by Chirikov et al. [11]. This is surprising because Anderson's result is concerned with the asymptotics for $n\to\infty$ of eigenfunctions of q.e.; here, we find very good exponential decay even for small n and for a nonstationary state. Evidently some mechanism must be working here that requires further investigation.

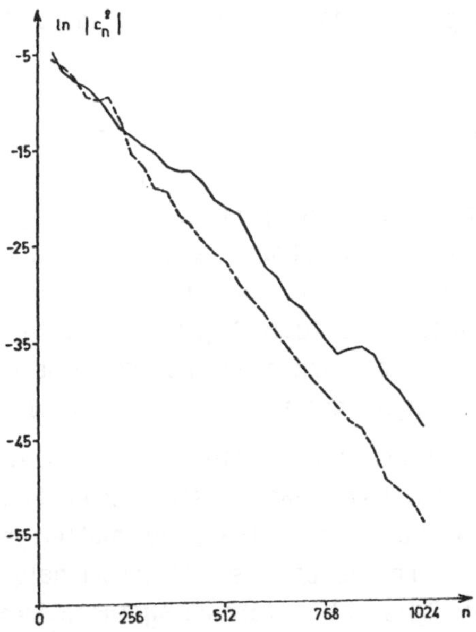

Fig. 2 - Distribution function $|c_n|^2$ after 50000 iterations of the quantum mapping for the case k=10 and τ=[0,25,1,1,1,....] (dashed line), τ=[0,25,100,200,400,800,.....] (solid line). Notice the fairly good exponential localization in momentum space over several orders of magnitude.

The above described behaviour holds for a typical irrational τ. For rational values of the kicked period τ, the rotator energy resonantly increases asymptotically as t^2. This is a strictly quantum phenomenon and does not occur in the corresponding classical system. This quantum peculiarity arises because the unperturbed quantum motion, as opposed to the classical, has the same period independent of initial state. In the resonant case we have specific information on the q.e. spectrum. In particular, it has been shown that, at resonance, the q.e. spectrum has an absolutely continuous component. Moreover, it has been shown that the growth of rotator energy with time t is proportional to t^2. But perhaps the most remarkable feature of resonance is that its quadratic growth of energy is the fastest possible [10].

In conclusion, for rational τ, the q.e. spectrum of the kicked rotator is continuous while for irrational τ the spectrum appears to be pure point. This situation is remindfull of the familiar process of destruction of invariant KAM tori in classical mechanics: rational tori are destroyed by the pertubation together with irrational-very close to rational-tori; morever as the perturbation strength increases, more and more tori are gradually destroyed.

One may therefore wonder whether a similar picture takes place for the quantum motion. In our case we know that in the absence of perturbation (ω=0) the q.e. spectrum is pure point; moreover, for any small perturbation and for rational τ the q.e. spectrum is continuous. What happens for irrational-very close to rational - τ? The analysis in ref. 13 shows that for $\tau=p/q$ the q.e. spectrum is made, typically, by q bands and that in the limit $p \to \infty$, $q \to \infty$ the width Δs of each band appears to be exponentially small ($\Delta s \sim (k/q)^q$). This would imply that the total width of all the q bands tends to zero as $q \to \infty$. Therefore for irrational τ the q.e. spectrum can be discrete or continuous with zero measure (singular continuous).
A recent analysis [14] has shown that, no matter how small is the perturbation, there is a non empty set of irrationals τ for which the spectrum is continuous. This set is formed by those irrationals which are very well approximated by rationals. We do not know how large is this set and in particular the dependence of its size from the perturbation strength. It may happen that the size of the set of irrationals τ yielding continuous spectrum increases with the perturbation strength or, on the contrary, it is possible that for a typical irrational τ, the spectrum remains pure point no matter how large is the perturbation. To our knowledge this is still an open question.

At this point one may turn to numerical experiments in order to see whether the nature of the q.e. spectrum and its dependence from the system parameters can be empirically detected. In Fig. 3 we present results of a typical experiment with irrational τ, using an initial δ-function excitation with k=10, K>1. The energy undergoes a diffusive like growth proportional to the break time t_b followed by a seemingly stationary plateau of small oscillations. However, since the energy must eventually become unbounded (continuous spectrum) or else exhibit a recurrent near return to initial state (point spectrum), the energy must eventually move away, either up or down, from the value shown in Fig. 3. Unfortunately, even though we have pushed the integration time interval near the limit of currently available super computers, we have been unable to decide which alternative actually occurs in this case.

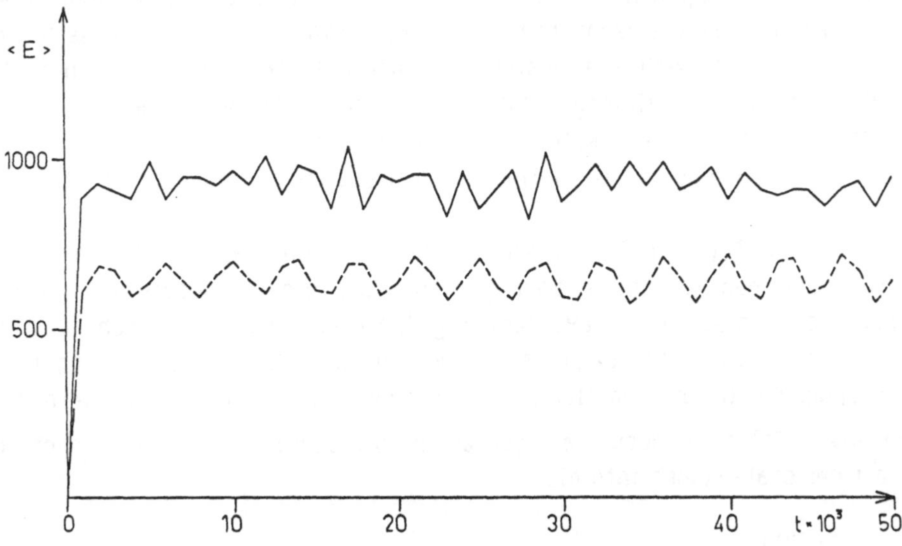

Fig. 3 - Average energy <E> versus time for the quantum kicked rotator with k=10 and τ=[0,25,1,1,1,....] (dashed line), τ=[0,25,100,200,400, 800,..] (solid line). Integration up to 50000 interations of the quantum mapping reveals that after the initial increase, the average energy oscillates around a fixed value .

Recently Dorizzi et al. [15] sought regularities in the q.e. spectrum by approximating the golden mean for τ by ratios of succesive Fibonacci numbers. They found that at each rational approximation of τ, the spectrum shows a new pattern and there is no indication of a tendency to a Cantor set. However the golden mean is a very "strong" irrational τ and in this case the continuous spectrum may be even lacking. Moreover it has been proven [16] that the spectrum for irrational τ has no gaps. This latter result excludes the possibility that the q.e. spectrum may be a Cantor set but does not preclude the possibility of singular continuous spectrum.

Numerical experiments which may suggest a singular continuous nature of the q.e. spectrum are shown in Fig. 4 were we plot the energy <E> versus time for several close values of rational τ. One can clearly see the resonant behaviour of <E(t)> corresponding to τ=[0,1] and to τ=[0,1,70]. In the first case, for irrationals of type [0,1,70,a_4,...], a parting of the ways from [0,1] occurs rather early, but then <E(t)> share a more or less extended phase of quadratic increase with the resonance [0,1,70], depending on the choice of a_4. Whatever choice of a_4 one may take, <E(t)> would then eventually display the resonance $[0,1,70,a_4]$; however, it is not possible to reach this point with the available computers. This indicates that the duration of the oscillatory "plateau" between subsequent resonances is rapidly increasing. The results in Fig. 4 are a sample of the behaviour of <E(t)> corresponding to the continuous irrational or nonresonant spectrum.

Therefore, in light of Fig. 4 and of the discussion above, the following picture emerges for the energy growth. This growth occurs by "jumps" separated by plateaus of exponentially increasing duration. Each jump is associated with the bandlike structure of the resonant spectrum corresponding to one particular convergent τ_n, and transition into the subsequent "plateau" occurs as soon as the structure of the actual spectrum on a finer scale comes into play.

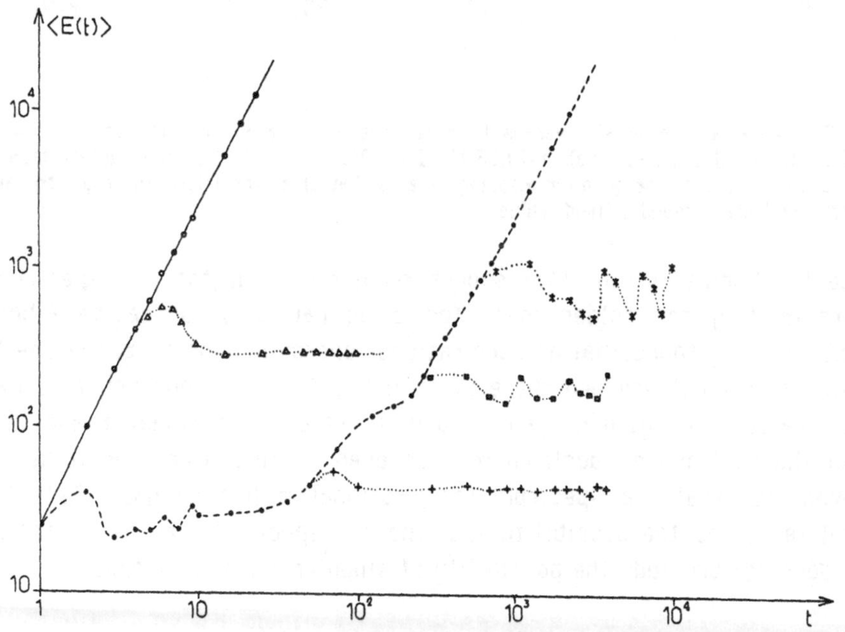

Fig. 4 - Average energy (E) versus time for the case k=10 and for several values of τ (o)τ= [0,1]; (Δ)τ=[0,1,1000]; ($\bullet$)τ = [0,1,70]; (*)τ = [0,1,70,10000]; ($\square$)τ = [0,1,70,1000]; (+)τ=[0,1,70,50]. The solid straight line represents the analytical result <E(t)> $=(k^2/4)t^2$ for the case τ=1. Notice also the asymptotic t^2 increase for the resonant case τ=[0,1,70] (dashed line).

The above considerations show also the possibility that an appropriate choice of the a_n in the continued fraction expansion $\tau \equiv [a_1, a_2, a_3, ...]$ may yield and energy growth which avoids both the quadratic increase as well as the plateau. Indeed in Fig. 5 we present a choice of τ which yield an energy growth up to $t \sim 2500$ even though the small value of quantum k=10 typically yield a break-time $t_b \sim k^2 = 100$.

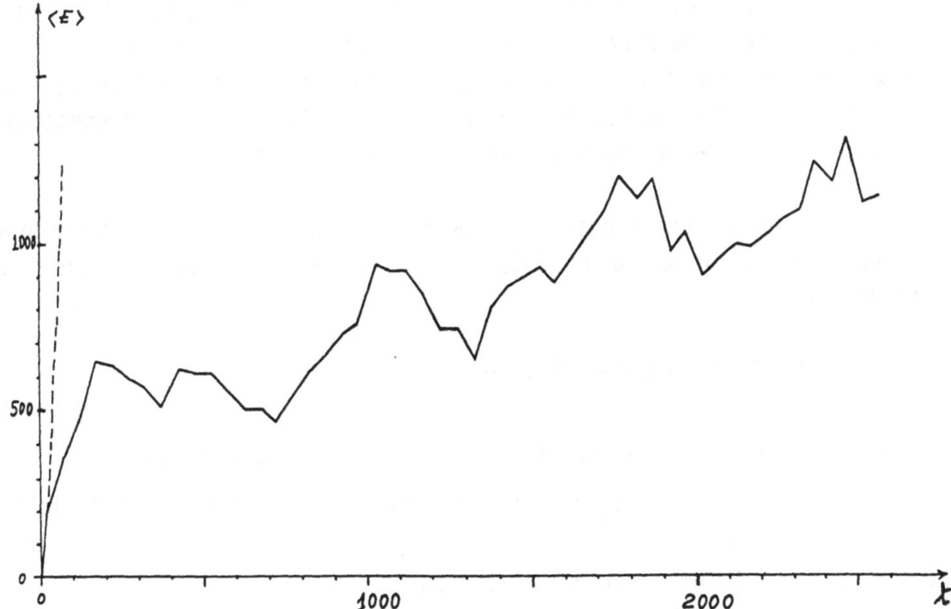

Fig. 5 - Average energy <E> versus time for k=10 τ=[0,1,5,10] (dotted line) and τ=[0,1,5,10,10,10,20] (full line). Notice the very large break time, $t_b \approx 2500$ for this latter case.

In conclusion we remark that even for those τ for which the spectrum is singular continuous thus yielding an indefinite average energy growth, the quantum system is, according to classical ergodic theory, no more chaotic then a weak mixing system.

The doubt has been rised that the quantum suppression of classical diffusion may be due to the presence of islands of stability in the classical motion. We would like to stress that this is not the case; indeed, as we have numerically verified, roughly the same quantum limitation of diffusion takes place in systems very close to the kicked rotator but which are completely chaotic without any stable region.

At this point one may be tempted to speculate that such limitation to quantum chaotic motion is inherent to the structure of quantum mechanics and therefore may be present also in systems subject to aperiodic, or even random, perturbations. The study of this latter class of systems seems very appropriate since it is within this class that one may get acquainted with the features of "quantum stochasticity" to be used for comparison when looking for the appearance of quantum dynamical stochasticity. In this respect the analysis [17] of a model of a randomly driven quantum system has shown that the quantum motion has an absolutely continuous spectrum in perfect analogy with the corresponding classical model.

In the same vein a quantum rotator in which kicks occur at random instants has been analyzed [18]. This model is described by the Hamiltonian:

$$H = d^2/d\theta^2 + \epsilon \cos\theta \sum_i \delta(t-t_i)$$

where t_i are random variables which define a Poisson process i.e. $\tau_1=t_1$, $\tau_2=t_2-t_1$, ...$\tau_n = t_{n+1}- t_n$ are independent random variables distributed according to a Poisson (exponential) law. It has been proved [18] that the kinetic energy of this quantum rotator is unbounded.

Finally the effect of noise on the quantum behaviour has been studied[19] and it has been found that, the presence of a small noise can have a strong effect on the quantum motion of a system which is classically chaotic and, in the semiclassical region, leads to a quantum diffusion.

In conclusion, the main indication which emerges from the analysis of the quantum δ-kicked rotator is that quantum mechanics places strong limitations to the classical chaotic motion. This limitation is certainly enhanced by the presence of cantori or islands of stability in the classical motion. Yet, quantum interference effects appear to suppress the chaotic diffusion process even if the classical motion is completely chaotic. This phenomenon is at least partially understood in terms of its connection with Anderson localization which is familiar in solid state physics. However this cannot be the all story since other models exhibit situations in which the quantum motion appear to follow the classical diffusion process [20,21].

The study of the δ-kicked rotator model provided us a great insight in the nature of quantum motion when the corresponding classical limit is chaotic. Many important details have still to be understood. Yet the limitation of chaos introduced by quantum mechanics seems to be a general, important phenomenon.

REFERENCES

1. For a clear and readable discussion see: Moser J., 1968, Memoirs Am. Math. Soc., No 81 .

2. Chirikov B.V., 1979, Phys. Rep. 52 263.

3. Ford J, Walker G.H., 1969, Phys. Rev. 188 416.

4. Henon H., Heiles C., 1964, Astron. J. 69 73.

5. Ford J., April 1983, Physics Today, For a popular discussion of this point see.

6. Casati G., Guarneri I. and Vivaldi F., 1983, Phys. Rev. Lett. 51 727.

7. Percival I.C., 1973, J. Phys. B L. 229.

8. T. Seligman Ed., 1986, Proceedings of the II International Conference on Quantum Chaos.

9. G. Casati, B.V. Chirikov, F.M. Izrailev and J. Ford, 1979, Lectures Notes in Physics Vol. 93.

10. G. Casati, J. Ford, F. Vivaldi, 1986, Phys. Rev. A 34, 1413.

11. B.V. Chirikov, F.M. Izarailev, and D.L. Shepelyansky, (1981) Sov. Sci. Rev. Sec. C 2, 209.

12. S. Fiscman, D.R. Grempel and R.E. Prange, 1982, Phys. Rev. Lett. 49, 509; and 1984, Phys. Rev. A 29, 1639.

13. F.M. Izrailev and D.L. Shepelyansky, 1980, Teor. Mat. Fiz. 43 , 417.

14. G. Casati and I. Guarneri, 1984, Commun. Math. Phys. 95, 121.

15. B. Dorizzi, B. Grammaticos, and Y. Pomeau, 1984, J. Stat. Phys. 37, 93.

16. J. Bellissard, 1985, Trends and developments in the eighties, Eds: S. Albeverio and Ph. Blanchard, World Scientific.

17. G. Casati and I Guarneri, 1983, Phys. Rev. Lett. 50, 640.

18. I. Guarneri, 1984, Lett. Nuovo Cimento 40, 171.

19. E. Ott, T. M. Antonsen, Jr., J.D. Hanson, 1984, Phys. Rev. Lett. 53, 2187.

FRACTALS IN QUANTUM MECHANICS!?

B ECKHARDT

1. INTRODUCTION

A striking manifestation of the complexity of motion in non-integrable Hamiltonian systems is the appearance of selfsimilar or fractal sets in phase space. Taking these fractals as one indicator for chaos, I will here discuss the possibility of "quantum chaos".

I begin with a short presentation of the various forms of fractals in classical mechanics. Using coherent states, I will then show that quantum mechanics is much smoother than classical mechanics. In particular, the smallest scales (smaller than h^N) in phase space are smeared out and fractals survive as a transient phenomenon only. In next section, I will address the more detailed question on the connection between classical periodic orbits and quantal eigenstates. The results for linear maps on a torus suggest that in addition to being smoothed out, many periodic orbits contribute to the eigenfunctions.

2. FRACTALS IN CLASSICAL MECHANICS

Presently, there are five forms of fractals known in classical mechanics. The first three are most conveniently studied via scaling properties of approximating periodic orbits: period-doubling attractor, KAM curve at criticality and cantori.

The period doubling attractor is defined in analogy to the dissipative case: If a periodic becomes unstable and if the eigenvalues of the linearization leave the unit circle through -1, then another orbit of twice the period becomes stable (tangent bifurcation). If this process continues ad infinitum, then the parameter values at which the bifurcations take place accumulate geometrically and at the limiting point, an attractor of period 2^∞, becomes stable. This mechanism has been studied by Greene et al (1981) for (conservative) two-dimensional maps. More recently, Mao, Satija and Hu (1985) have shown that for four-dimensional maps, there is also the possibility of a different scaling behaviour.

It is well known that according to the KAM theorem, tori with rational winding numbers are most susceptible to perturbations. On the other hand, tori with irrational winding numbers are rather robust. But if the perturbations get too strong, they break. Numerical results by Greene (1979), Shenker and Kadanoff (1982) and MacKay (1983) indicate that irrational tori at criticality are selfsimilar smooth but nowhere differentiable objects in phase space. Above criticality, a cantorus remains (MacKay, Meiss and Percival, 1984; Li and Bak, 1986). An important difference is that at criticality, the torus is still stable, but the cantorus is unstable. This has consequences for the observability of the scaling to be described below.

A fourth example of fractals was introduced by Umberger and Farmer (1985). They studied the phase space volume occupied by a single orbit initiated in a stochastic layer. Now it is well known that in the stochastic layer there are islands, surrounded by islands, ..., ad infinitum. Even if the boundaries between islands and layer were smooth (and there is evidence that they are not, see Greene, MacKay and Stark, 1986), there is a fractal induced by the clustering of islands. This shows up e.g. if one estimates the phase space volume of the layer by covering it with cubes of area ϵ. Umberger and Farmer find $A(\epsilon) = A(o) + \text{const}\ \epsilon^{\beta}$ with $\beta \neq 1$, suggesting that the stochastic layer is a fat fractal.

Our final example is taken from scattering theory. It has been known for some time that in a special case of the three-body problem, chaotic scattering is possible (Sitnikov's problem, see Moser, 1973). Consequences for the scattering data have only recently been studied in an example from hydrodynamics (four vortex motion, Manakov and Shchur, 1983; Eckhardt and Aref, 1986), in reactive scattering (Noid, Gray and Rice, 1986) and in a model in potential scattering (Eckhardt and Jung, 1986). For definiteness, consider scattering off a Henon Heiles type potential with exponential cut-off to define the asymptotics. Fix the in-going direction and monitor the angle of deflection as a function of impact parameter. For most values, the scattering angle will vary smoothly. However, in some regions, it oscillates wildly, as shown in Fig. 1. Magnification of the irregular regions indicates that they persist on all scales. We conjecture that there is a Cantor set of values for the impact parameter for which asymptotic trapping occurs. For more details, see the above cited literature.

3. THE UNCERTAINTY RELATIONS AND CONSEQUENCES

Crucial for all fractals in classical mechanics is the lack of a smallest scale in phase space which allows self-similarity to persist on all scales. The Heisenberg uncertainty relationship $\Delta p \Delta q \gtrsim h$ certainly introduces a smallest scale in phase space for quantum mechanics. A simple way of implementing this finite resolution in classical mechanics is by smearing out any density $\rho(p,q)$ e.g. by Gaussians of width σ_p and σ_q,

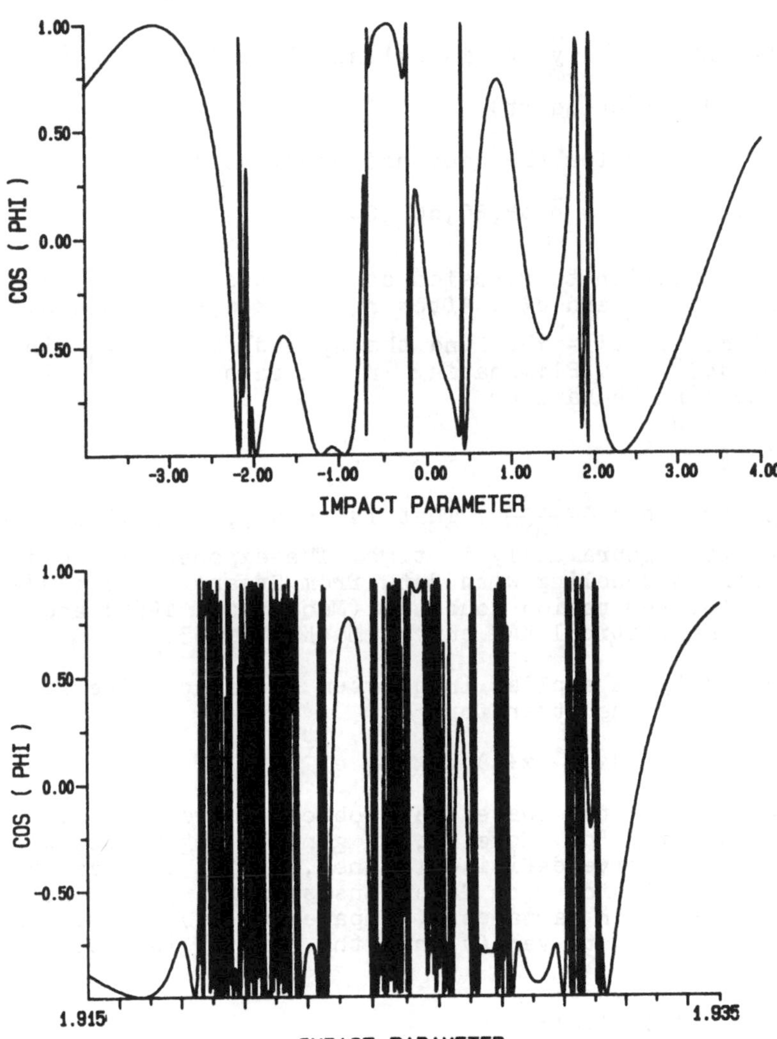

Fig. 1. The cosine of the angle of deflection as a function of impact parameter for a system that shows irregular scattering. The bottom figure shows a magnification near b = 1.925 . For details, see (Eckhardt and Jung, 1986).

$$\bar{\rho}(P,Q) = \text{const} \int dpdq\ \rho(p,q) \exp\left(-\frac{(q-Q)^2}{2\sigma_q^2} - \frac{(p-P)^2}{2\sigma_p^2}\right) \quad . \tag{1}$$

If the "point" density ρ is selfsimilar, i.e.

$$\rho(p,q,t) = \rho(\alpha p,\beta q,\gamma t) \tag{2}$$

then this implies for the smoothed density $\bar{\rho}$

$$\bar{\rho}(P,Q,\sigma_p,\sigma_q,t) = \bar{\rho}(\alpha P,\beta Q,\alpha\sigma_p,\beta\sigma_q,\gamma t) \quad , \tag{3}$$

so that in addition to rescaled coordinates and time, one has new variances σ_p' and σ_q' . After n iterations of this process, the new time is $t' = \gamma^n t$ and the spread $S' = \sigma_p'\sigma_q'$ in phase is $S' = (\alpha\beta)^n S_o$. Eliminating n in favour of time, one finds a scaling behaviour

$$S' = \left(\frac{t'}{t}\right)^{\sigma} S_o \tag{4}$$

with exponent $\sigma = \frac{\ell n(\alpha\beta)}{\ell n\gamma}$. That is to say, an initial uncertainty grows algebraically in time. The exponent is $\sigma = 6.03$ for 2-D period doubling with data from Greene et al (1981), $\sigma = 6.93$ for 4-D period doubling (Mao et al, 1985) and $\sigma = 3.05$ for the critical KAM curve (MacKay, 1983).

A similar analysis applies in quantum mechanics. One first identifies the Wigner transform

$$W(p,q) = \frac{1}{\pi h} \int dx\ \psi^*(x+q)\ \psi(x-q)\ e^{2ipx/h} \tag{5}$$

of a state ψ as the phase space object corresponding to the classical density ρ . However, by general arguments, W(p,q) need not be positive definite (Wigner, 1971). This can be repaired by convolution with Gaussians satisfying the uncertainty relations for their variances (compare Eq. (1)). It is then a simple computation to verify that the new $\overline{W}$ is positive definite

$$\overline{W}(p,q) = |\langle\phi_{pq}|\psi\rangle|^2 \tag{6}$$

with

$$\phi_{pq}(x) = (2\pi\sigma_q^2)^{-1/4} \exp\left(-\frac{(x-q)^2}{4\sigma_q^2} + i\ \frac{px}{h}\right) \tag{7}$$

and $\sigma_p = \frac{h}{4\sigma_q}$ (so that ϕ_{pq} is a minimal uncertainty wave packet). The representation (6),(7) is also known as coherent state representation of the Wigner density. The equations of motion for the smoothed $\overline{W}$ and the smoothed classical density $\bar{\rho}$ are the same up to order $\Theta(h^2)$. Therefore, the leading order h scaling in quantum mechanics is the same as that of finite resolution corrections in classical mechanics

$$h(t) \sim t^{\sigma}\, h(o) \quad . \tag{8}$$

Since there is an algebraic time-dependence, this process will dominate linear ones like spreading of wave packets positioned on classically stable objects (like period doubling attractors or critical KAM curve), but it will be concealed by exponential processes like the spreading of wave packets on unstable structures (like cantori).

The argument presented here is the coherent state version of the stationary phase analysis of Grempel, Fishman and Prange (1984).

4. EIGENFUNCTIONS

I now want to make a few remarks on fractals in eigenfunctions, i.e. on the connection between classical invariant objects and similar structures in eigenfunctions. Note that fractal eigenstates of tight binding Hamiltonians (Aoki, 1982, 1986; Soukoulis and Economou, 1984; Roman, 1986) seem to be a quantum interference phenomenon without classical counterpart.

In the preceding section, we saw that the smallest scales of fractals are eliminated upon quantization. Here we will see that several fractals contribute to eigenfunctions, thus even further blurring the self-similarity. It is usually very difficult to get detailed information about eigenstates of nonintegrable systems in the semiclassical limit (see Voros, 1976, 1977, 1979; Berry, 1977), so I will confine my discussion to linear maps on the torus, a system that is completely soluble (Hannay and Berry, 1980; Eckhardt, 1986). Moreover, I will talk about periodic orbits only, but since the fractal behaviour may be thought of as asymptotic scaling of approximating periodic orbits, I loose no generality.

Consider the mapping

$$\begin{pmatrix} p_{n+1} \\ q_{n+1} \end{pmatrix} = \begin{pmatrix} a & b \\ c & d \end{pmatrix} \begin{pmatrix} p_n \\ q_n \end{pmatrix} \bmod 1 \tag{9}$$

with a,b,c,d integer and $\det \begin{pmatrix} a & b \\ c & d \end{pmatrix} = 1$, thus mapping the unit square of R^2 onto itself. Since both positions and momenta are periodic, the quantum system lives on a finite square grid in phase space. If there are N points in position (i.e. $q = k/n$ with $k=0,1,\ldots,N-1$) then the effective Planck's constant is $h=1/N$. Therefore, the semiclassical limit is also the limit $N \to \infty$.

Quantum states are given by (normalized) complex vectors (ψ_k) $k=0,1,\ldots,N-1$. The unitary propagator U for one time step is given by a unitary $N\times N$ matrix and typically has the form

$$U_{k\ell} = \exp \frac{i\pi}{N} (Ak^2 + B\ell^2 + Ck\ell) \tag{10}$$

where the integer coefficients A,B,C depend on the matrix elements. Of course, the quantum dynamics is most conveniently described in terms of eigenvalues and eigenvectors of U . It can be shown (Hannay and Berry, 1980) that the eigenvalues are roots of unity

$$\chi \in \left\{ e^{2\pi i\ (\ell/n(N))} \mid \ell=0,1,\ldots,n(N)-1 \right\}$$

where $n(N)$ is determined from $T^{n(N)} = 1 \bmod N$, i.e. $n(N)$ is the period of the mapping T confined to a finite grid. Until recently, only numerical results on $n(N)$ were available indicating a strong dependence on number theoretical properties of N . It was found that for most N prime, $n(N)$ divides $N+1$ or $N-1$. In this case, eigenvalues are least degenerate and typical behaviour of eigenvectors can be studied. These numerical results were put on firm grounds by Percival and Vivaldi (1986) using ideal theory (see also Vivaldi's contribution to this volume).

As shown in (Eckhardt, 1986), the eigenfunctions can be represented in closed form as

$$\psi_k = \sum_\mu \exp\left(\frac{i\pi}{N} q_\mu k^2 + i\delta_\mu\right) \tag{11}$$

with integers q_μ and phases δ_μ determined from recursion relations. The dependence on the eigenvalues is hidden in the phases. The number of elements in the sum turns out to be $n(N)$, $n(N)/2$ or 1 . As I will show now, this is connected with periodic orbits.

The classical map is linear and, therefore, it also determines the time evolution of the Wigner density. Since the Wigner density of eigenstates is invariant, it follows that it is a superposition of densities concentrated along classical periodic orbits. Projection of the Wigner density onto coordinate space yields, by construction, the eigenfunctions. An explanation for the number of elements in (11) then is that periodic orbits group according to degeneracies of eigenvalues and symmetries of the map, but within each class, all orbits contribute to the eigenfunctions.

This shows that there is no simple, much less a one-to-one link between classical periodic orbits and quantum eigenstates.

5. CONCLUSIONS

The preceding discussion shows that quantum mechanics is not as singular as classical mechanics. The Heisenberg uncertainty relations introduce an inner, smallest scale in quantum mechanics, below which all densities are smooth. Similar reasoning applies to the fat fractal stochastic layer (here the result of Brossard and Carmona, 1986 is relevant). And finally, an inner scale in irregular scattering is introduced by the spreading of the wave packet and/or tunneling. Quantum mechanically, it is impossible for a wave packet to be trapped indefinitely in this potential.

To summarize: Classical fractals persist in quantum mechanics only as transient phenomena. An important open problem are the next to leading order h corrections.

The completely soluble model of linear maps on a torus shows another effect of quantization, visible in eigenstates: Not only are classical fractals spread out, they are also mixed to form quantal eigenstates. This is a simple illustration of the conclusion by Helton and Tabor (1985) that eigenfunctions are concentrated on convex linear combinations of periodic orbits.

REFERENCES

Aoki, H. 1983, J. Phys. C: Solid State Phys. 16 L205-L208
Aoki, H. 1986, Phys. Rev. B 33 7310-7313
Berry, M.V. 1977, J. Phys. A: Math. Gen. 12 2083-2091
Brossard, J. and Carmona, R. 1986, Commun. Math. Phys. 104 103-122
Eckhardt, B. 1986, J. Phys. A: Math. Gen. 19 1823-1831
Eckhardt, B. and Aref, M. 1986, Collision dynamics of vortex pairs, in preparation
Eckhardt, B. and Jung, C. 1986, Regular and irregular potential scattering, to appear in J. Phys. A: Math. Gen.
Greene, J.M. 1979, J. Math. Phys. 20 1183-1201
Greene, J.M., MacKay, R.S., Vivaldi, F. and Feigenbaum, M.J. 1981, Physica 3D 468-486
Grempel, D.R., Fishman, S. and Prange, R.E. 1984, Phys. Rev. Lett. 53 1212-1216
Hannay, J.M. and Berry, M.B. 1980, Physica 1D 267-290
Helton, J.W. and Tabor, M. 1985, Physica 14D 409-415
Li, A. and Bak, P. 1986, Phys. Rev. Lett. 57 655-658
MacKay, R.S. 1983, Physica 7D 283-300
MacKay, R.S., Meiss, J.D. and Percival, I.C. 1984, Physica 13D 55-81
Manakov, S.V. and Shchur, L.N. 1983, Sov. Phys. JETP Lett. 37 54-57
Mao, J.M., Satija, I.I. and Hu, B. 1985, Phys. Rev. A32, 1927-1929
Moser, J. 1973, Stable and Random Motions in Dynamical Systems, Princeton University Press
Noid, D.W., Gray, S.K. and Rice S.A. 1986, J. Chem. Phys. 84 2649-2652
Percival, I.C. and Vivaldi, F. 1986, Arithmetical properties of strongly chaotic motion, to appear in Physica D

Roman, H.E. 1986, J. Phys. C: Solid State Phys. 19 L285-L288
Shenker, S.J. and Kadanoff, L.P. 1982, J. Stat. Phys. 27 631-655
Soukoulis, C.M. and Economou, E.N. 1984, Phys. Rev. Lett. 52 565-568
Umberger, D.K. and Farmer, J.D. 1985, Phys. Rev. Lett. 55 661-664
Voros, A. 1976, Ann. Inst. H. Poinc. A24 31-90
Voros, A. 1977, Ann. Inst. H. Poinc. A26, 343-403
Voros, A. 1979, in Stochastic Behaviour in Classical and Quantum Hamiltonian Systems, eds. G. Casati and J. Ford, Springer, Berlin, pp 326-333
Wigner, E.P. 1971, in Perspectives in Quantum Theory, eds. W. Yourgran and A. van der Merwe, MIT Press, Cambridge, pp 25-36

ERGODIC SEMICLASSICAL QUANTUM MECHANICS

M FEINGOLD

1. INTRODUCTION

After major progress in the last years, chaotic phenomena in classical systems are fairly well understood (Lichtenberg and Lieberman, 1983). On the other hand, many questions regarding the behavior of the quantized version of chaotic systems are still open (Berry, 1983). Moreover, the predictive ability in the field of quantum chaos is mostly qualitative. For time independent Hamiltonian systems, the few quantitative results are based either on semiclassical arguments (Berry, 1977; Berry and Tabor, 1977; Gutzwiller, 1967, 1969, 1970, 1971 1980; Pechukas, 1983) or on the random matrix theory (Mehta, 1967; Brody, Flores, French, Mello, Pandey and Wong, 1981). Regarding the former, we have no working scheme by which non-integrable systems can be semiclassically quantized. The Einstein-Brillouin-Keller (van Vleck, 1928; Keller, 1958) method is bound to break down in those regions of phase space in which the invariant manifolds are no longer tori. As to the latter, it is not clear yet to what extent random matrix theory applies to quantum chaos (Berry, 1985; Gelbart, Rice and Freed, 1972). The Gaussian ensembles have the same probability distribution for all the off-diagonal matrix elements. This homogeneity is not characteristic for a Hamiltonian matrix in some generic representation.

It is the purpose of this paper to argue that rather a combination of the two methods, to be called ergodic quantization from now on, will render good quantitative predictions in the field of quantum chaos. We will start by shortly reminding the reader how ergodic quantization works for the well known case of the spectrum. Next, an analogous treatment will be developed for the matrix elements of operators in the energy representation (the representation in which the Hamiltonian is diagonal).

Ergodic quantization is done in two steps. First, the mean behavior of the quantum object (spectrum, matrix elements, etc) is semiclassically calculated. This is, loosely speaking, the lowest term in the $\hbar$ expansion. This step generates the overall structure which is missing in the random matrices. Second, normalizing by the mean behavior, the "unfolded" object will be structureless, up to random fluctuations. Naturally enough, it is assumed that these fluctuations are well described by the Gaussian ensembles. Our arguments are limited to the strongly chaotic regime, partly because we will be using the ergodic assumption.

Moreover, we believe that the classically chaotic motion is responsible for the randomization of the fluctuations.

For the spectrum this procedure is extremely simple. The semiclassical approximation of the density of states

$$\rho(E) = \sum_i \delta(E - E_i) \tag{1.1}$$

is obtained by counting how many hypercubes with side h of the same dimensionality as the configuration space, can fit into the energy surface

$$\rho_{SC}(E) = \frac{1}{h^N} \iint d\vec{q}\,d\vec{p}\ \delta[E - H(\vec{q},\vec{p})] \tag{1.2}$$

Using ρ_{SC}, the unfolded spectrum, $\tilde{E}_i$, is generated

$$\tilde{E}_i = E_i \rho_{SC}(E_i) \tag{1.3}$$

As schematically shown in Fig. 1, the $\tilde{E}$ spectrum has lost the overall structure which is obvious in the original E spectrum. The mean spacing, $\tilde{D}$, between two neighboring levels of the $\tilde{E}$ spectrum is unity. On the other hand, the fluctuations of the unfolded spectrum are random and their statistical properties are well described by the Gaussian ensembles predictions. Specifically, for time reversal invariant systems, the Hamiltonian is real and therefore the Gaussian Orthogonal Ensemble (GOE) is appropriate. The spacing, $s = \tilde{E}_{i+1} - \tilde{E}_i$ is predicted by the GOE to have a probability distribution which is well approximated by the Wigner result (Mehta, 1960; Gaudin, 1961)

Fig. 1. The effect of the unfolding procedure for the spectrum (schematic illustration).

$$P(s) = \frac{\pi s}{2\tilde{D}^2} \exp\left[-\frac{\pi s^2}{4\tilde{D}^2}\right] \tag{1.4}$$

Lately, (1.4) has been numerically checked for several chaotic Hamiltonian models; good agreement has been found with increasing statistical significance (McDonald and Kaufman, 1979; Casati, Valz-Gris and Guarneri, 1980; Bohigas, Giannoni and Schmidt, 1984; Haller, Koppel and Cederbaum, 1984; Seligman, Verbaarschot and Zirnbauer, 1984).

From the above analysis of chaotic spectra, it might seem that the ergodic quantization scheme is just ready to be generalized to any other quantum objects. However, for the sake of completeness, we should mention that, when the Δ_3 statistic is employed on the unfolded spectrum, a new problem arises. The $\Delta_3(r)$ is defined (Dyson and Mehta, 1963) as

$$\Delta_3(r) = < \min_{A,B} \frac{1}{2L} \int_0^{2L} [N(\tilde{E}) - A\tilde{E} - B]^2 \, d\tilde{E} > \tag{1.5}$$

where $r = 2L/D$ is the mean number of energy levels in the $(0,2L)$ interval, <...> denotes spectral or ensemble averaging and $N(\tilde{E})$ is the integrated density of states

$$N(\tilde{E}) = \sum_i \theta(\tilde{E} - \tilde{E}_i) \tag{1.6}$$

Therefore, Δ_3 measures the fluctuation in the density of states. For the GOE

$$\Delta_3(r) = \frac{1}{\pi^2}[\ln r - 0.0687] \tag{1.7}$$

It was recently pointed out (Berry, 1985) that for chaotic systems, $\Delta_3(r)$ will follow (1.7) only for small r, $r<r^*$, where $r^* \propto \ln(h^{-1})$; for $r>r^*$ its behavior will saturate in a non-universal fashion which will in general differ from (1.7).

Because the $\Delta_3(r)$ is mainly a two-point function of the spectral sequence, we reach the admittedly heuristic conclusion that the ergodic quantization procedure is applicable only to local quantum objects. In the present discussion locality will denote a small energy range.

2. ERGODIC QUANTIZATION FOR MATRIX ELEMENTS - THEORETICAL PREDICTIONS

2.1 The Semiclassical Step

By analogy with the spacings distribution analysis just reviewed in the previous section, we will develop in this section some methods by which semiclassical information about matrix elements of generic operators in the energy representation can be obtained. This is the first step of the ergodic quantization for matrix elements, exactly as finding ρ_{SC} was the first step of the same procedure for the spectrum (Feingold, Moiseyev and Peres, 1985; Feingold and Peres, 1986).

Suppose A is our operator. Then in the Heisenberg picture

$$A_{jk}(t) = A_{jk}(t{=}0)\ exp\,[2\pi i\,(E_j - E_k)t/\hbar] \tag{2.1}$$

and if we time-average (2.1)

$$\overline{A}_{jk} \equiv \lim_{T\to\infty} \int_0^T A_{jk}(t)\,dt = \begin{cases} A_{jk}(t{=}0) & E_j = E_k \\ 0 & E_j \neq E_k \end{cases} \tag{2.2}$$

For a non-degenerate spectrum, $E_j = E_k$ implies $j{=}k$. In this case, equation (2.2) is an equality between the quantum time average of operator the A at energy E, $\overline{A}_{QU}(E)$, and the diagonal element at the same energy. Invoking the correspondence principle for $\hbar\to 0$ in the second relation of (2.3) and the ergodic assumption in the third, we obtain

$$<E\,|A\,|E> = \overline{A}_{QU}(E) \approx \overline{A}_{CL}(E) = \{A\}_E \tag{2.3}$$

where $\{...\}$ denotes microcanonical averaging and the approximative equality sign will become exact equality as the semiclassical limit is approached. The equality between the diagonal elements and the microcanonical average emerging from (2.3) is the main ingredient of the semiclassical step. As it is, (2.3) already gives an estimation of the matrix elements using only classical objects. However, this is still a rather weak prediction as it concerns only a small part of the matrix: its diagonal.

In order to obtain information about off-diagonal matrix elements, the following main relation will be used

$$M_{2n} \equiv \sum_k (E_j - E_k)^{2n} \, |A_{jk}|^2 = (i)^{2n} \langle E_j | ([H,[H,...[H,A]...]])^2 | E_j \rangle =$$
$$\approx (\hbar)^{2n} \{([H,[H,...[H,A]_{PB}...]_{PB}]_{PB})^2\}_{E_j} =$$
$$= (\hbar)^{2n} \{(\frac{d^n A}{dt^n})^2\}_{E_j} \tag{2.4}$$

where [...] denotes the quantum commutator while $[...]_{PB}$ is the Poisson bracket. In the first equality of (2.4), averages of the off-diagonal elements function are brought into the diagonal element form of another operator. Next, using (2.3), the diagonal element is estimated by the microcanonical average of the same operator. One should notice however, that while the quantum operator contains commutator brackets, in the classical one these are replaced by Poisson brackets. As a result of this replacement, a $(\hbar/i)^{2n}$ factor is obtained. By semiclassically estimating averages of the type appearing in (2.4) for different n-s we can get information about the semiclassical function $|A_{jk}|^2_{SC}$ which approximates the variation of the off-diagonal matrix elements. $|A_{jk}|^2_{SC}$ is a smoothed version of the exact variation of the matrix elements and the (2.4)-type averages are its n-order moments. The more moments we will calculate the more accuracy on the prediction of the $|A_{jk}|^2_{SC}$ function we will gain.

Let us now exemplify the use of Eq. (2.4). We define the energy width of the A_{jk} matrix, ΔE, as follows

$$(\Delta E)^2 = \frac{\sum_k |A_{jk}|^2 (E_j - E_k)^2}{\sum_k |A_{jk}|^2} \tag{2.5}$$

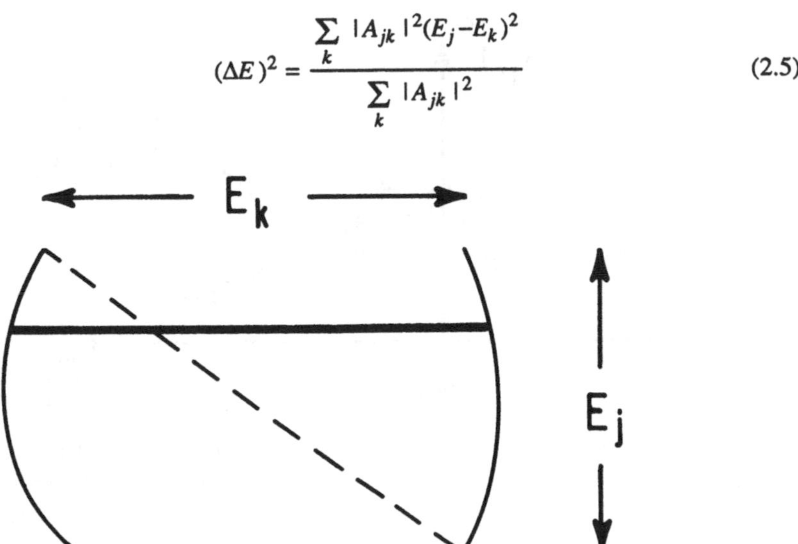

Fig. 2. One row of the operator matrix in the energy representation.

which is basically the second moment of the $|A_{jk}|^2(E)$ function (see Fig. 3). Using (2.4) and (2.3), ΔE can be semiclassically estimated

$$(\Delta E)^2 = \frac{<E_j|([H,A])^2|E_j>}{<E_j|A^2|E_j> - <E_j|A|E_j>^2} \approx \hbar^2 \frac{\{[H,A]_{PB}\}_{E_j}}{\{A^2\}_{E_j} - \{A\}^2_{E_j}} \tag{2.6}$$

where the last expression gives the second moment of $|A_{jk}|^2_{SC}$, ΔE_{SC}. Thus, as emphasized in Fig. 3, in the $\hbar \to 0$ limit, the A operator is represented by a band matrix (a matrix which has elements significantly differing from zero only in some band around the diagonal). Moreover, the width of the band in energy units scales like $\hbar$. It is interesting to notice that on the other hand, the band width in index units, $\Delta E/D$, scales like $\hbar^{1-N}$. For N=2 this leads to a $\hbar^{-1}$ divergence in the semiclassical limit. At this stage one would be tempted to conclude that in the $\hbar \to 0$ limit, the matrix will have all over elements of the same order of magnitude and therefore the random matrix assumptions will be appropriate in this regime whenever $N \geq 2$. However, at least for algebraic Hamiltonians (finite matrices), the dimension of the A matrix scales as $\hbar^{-N}$. Thus, in the semiclassical limit, the ratio between the band width and the matrix size goes to zero as $\hbar$ and a non-trivial variation in the matrix elements magnitude emerges. This variation has to be canceled in order to obtain the structureless random matrix feature. Once $|A_{jk}|^2_{SC}$ is known, the unfolded matrix $|\tilde{A}_{jk}|^2$ can be obtained

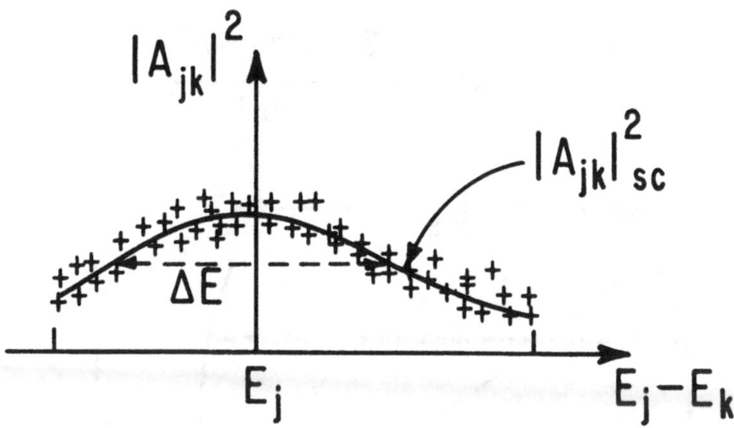

Fig. 3. The variation of the matrix elements in the row picked in Fig. 2 as a function of the distance from the diagonal in energy units. The small crosses denote the exact matrix elements while the smooth line is the semiclassical estimation, $|A_{jk}|^2_{SC}$.

$$|\tilde{A}_{jk}|^2 = \frac{|A_{jk}|^2}{|A_{jk}|^2_{SC}} \tag{2.7}$$

All matrix elements in $|\tilde{A}_{jk}|^2$, if regarded as random variables, have the same average which is equal to one. In the next section we will analyze the probability distribution of $|\tilde{A}_{jk}|^2$ using random matrix theory arguments. But before that, let us develop a more direct method to calculate $|A_{jk}|^2_{SC}$.

Similarly to (2.4), the second method is also based on a sum rule. That is

$$\begin{aligned}
&\sum_k exp\,[i(E_j - E_k)t/\hbar]\,|A_{jk}|^2 = \langle E_j|e^{iHt/\hbar}Ae^{-iHt/\hbar}A|E_j\rangle = \\
&= \langle E_j|A(t)A(0)|E_j\rangle \approx \{A(t)A(0)\}_{E_j} = \lim_{T\to\infty}\frac{1}{T}\int_0^T A(t+\tau)A(\tau)\,d\tau \equiv \\
&\equiv C(t) = \frac{1}{2\pi}\int_{-\infty}^{\infty} d\omega\, e^{-i\omega t}S(\omega)
\end{aligned} \tag{2.8}$$

where $C(t)$ is the autocorrelation function and $S(\omega)$ the power spectrum of the $A(t)$ dynamical variable. The third relation in (2.8) is based on Eq. (2.3) while the fourth is a result of the ergodic assumption. The last step in (2.8) is the well known Wiener-Khintchine theorem (Wiener, 1930; Khintchine, 1934). As in the case of Eq. (2.3), in (2.8) only the emerging relation between the first and the last expressions is used. In the semiclassical limit the spectrum becomes almost continuous such that

$$\sum_k \cdots \approx \int \cdots \rho(E)\, dE \tag{2.9}$$

In this case also, the almost equality sign will become exact equality when $\hbar\to 0$. Using (2.8), we obtain

$$\int dE\ |A_{jk}|^2(E)\,\rho(E)e^{-iEt/\hbar} \approx \frac{1}{2\pi}\int_{-\infty}^{\infty} d\omega\, e^{-i\omega t}S(\omega) \tag{2.10}$$

which implies

$$|A_{jk}|^2(E_j) \approx \frac{S\left[\dfrac{E_j - E_k}{\hbar}\right]}{2\pi\rho(E_j)} \equiv |A_{jk}|^2_{SC}(E_j) \tag{2.11}$$

where we have used $\omega = (E_j - E_k)/\hbar$. The (2.11) result is hardly surprising if one remembers that for regular systems, where the power spectrum is discrete, the matrix elements are given by $S(\omega)$ itself.

Finding $|A_{jk}|^2{}_{SC}$ and the (2.7) normalization procedure end up the semiclassical step in the ergodic quantization of the matrix elements. The $|\tilde{A}_{jk}|^2$ matrix we are left with after this step is structureless up to fluctuations. As in the case of the spectrum, we conjecture that the fluctuations are well described by the appropriate Gaussian random ensemble. The belief in the correctness of this conjecture is stimulated by the fact that the analogous conjecture for the spectrum was numerically found to be extremely accurate.

There are two features of the fluctuations one should keep in mind:

1) fluctuations are a result of the $\hbar$ finiteness and will disappear when $\hbar \to 0$,
2) the exact quantum matrix elements could be in principle found from an expansion in powers of $\hbar$.

The apparently random aspect of the fluctuations emerges from the high sensitivity of the $\hbar$-expansion on the index position in the matrix. Therefore, the random matrix approach to analyzing the fluctuations should be regarded as no more than a working approximation. Intuitively, the reason it works is that the underlying classical motion is chaotic. That gives rise to random fluctuations. However, the real reason for the applicability of the random matrix approach and its limitations can be found only in the $\hbar$-expansion.

2.2 The random matrix step

In the late fifties nuclear physicists started to look for a statistical approach to nuclear properties. Nuclei are very complicated systems with lots of particles, all strongly interacting among themselves. It is therefore reasonable to describe their Hamiltonian, H, in a generic basis, as a random matrix. The elements of a random matrix are random variables. Furthermore, every realization of the random matrix is regarded as a member in an ensemble. This is the way random matrix theory came about. We will limit ourselves to the case of the Gaussian Orthogonal Ensemble (GOE) which applies to problems with time-reversal symmetry for which H is real.

The assumptions for GOE are the following: if $H \in$ GOE then

1. H is a real, symmetric $M \times M$ matrix,
2. H_{ij} are independent random variables,
3. $P(H) = P(O^{-1}HO)$ where $O^{-1} = O^T$ is an orthogonal transformation matrix and because of assumption 2.

$$P(H) = \prod_{i<j} P_{ij}(H_{ij}) \tag{2.12}$$

is the joint probability distribution for all the matrix elements in H.

A number of important results, given in the order in which they have been obtained, are

I. the probability distribution of the H_{ij} is

$$P(H_{ij}) = \begin{cases} G(0,\sigma) & i \neq j \\ G(0,2\sigma) & i=j \end{cases} \tag{2.13}$$

where $G(0,\sigma)$ is a Gaussian distribution with zero mean and σ variance,

II. the joint probability distribution of the eigenvalues of H, E_i, is

$$P(E_1,E_2,\ldots,E_M) = K_M \prod_{i<j} |E_i-E_j| \; exp[-\frac{1}{4\sigma^2}\sum_j E^2{}_j] \tag{2.14}$$

III. by integrating (2.14) over all E_i such that $s=E_{j+1}-E_j$ stays constant and summing over j we get a nearest neighbors spacings distribution very close to (1.4).

These results are important for the spectral analysis. In order to find the distribution of the matrix elements of operators in the eigenvector basis we will first have to find the distribution of the eigenvector elements. Because we want the eigenvectors to be normalized, invariance of their distribution under orthogonal transformations (assumption 2.) will simply require that they be uniformly distributed on an unit sphere minimally embedded in the M dimensional space. The probability distribution for one vector element, a_i, will be

$$P(x=a_i) = \int_{-\infty}^{\infty} \cdots \int_{-\infty}^{\infty} da_1 da_2 \cdots da_{i-1} da_{i+1} \cdots da_M \; \delta(\sum_{i=1}^{M} a_i^2 - 1) =$$

$$= \frac{\Gamma\left(\frac{M}{2}\right)}{\sqrt{\pi}\,\Gamma\left(\frac{M-1}{2}\right)}(1-x^2) \underset{M\to\infty}{\rightarrow} \sqrt{\frac{M}{2\pi}} \exp\left(-\frac{Mx^2}{2}\right) \tag{2.15}$$

That is to say, the vector element has a $G(0,\frac{1}{\sqrt{M}})$ distribution for large M. The distribution of $a^2{}_i$ is obtained by a simple change of variables

$$P\,(y{=}a_i^2) = \frac{1}{\sqrt{2\pi y\,<y>}}\,\exp\left[-\frac{y}{2<y>}\right] \tag{2.16}$$

Eq. (2.16) is the well known Thomas-Porter distribution for the squared vector elements (Porter and Thomas, 1956; Heller and Sundberg, 1984). It diverges like $y^{-1/2}$ for small y and decays exponentially for large y. The average of a_i^2, $<a_i^2>$, is equal to $1/M$. This satisfies the normalization condition and is the same for all vector elements emphasizing the non-existence of any structure on the average.

The next and last step will be to analyze the matrix elements of operators in the eigenvector basis. Suppose all eigenvectors (and the Hamiltonian matrices as well) are written in a specific basis, $|\alpha_i>$, where i=1,...,M. Then one could always find an operator such that

$$x = <f\,|\alpha_i> = <f\,|A\,|i> \tag{2.17}$$

where $|f>$ and $|i>$ are eigenstates. Conversely, for every operator, A, there exists a basis $|\alpha_i>$ such that Eq. (2.17) will hold. Both directions of the (2.17) equality can be easily proven by construction. Therefore, matrix elements are also Gaussian distributed while their squares display a Thomas-Porter distribution. Because our semiclassical estimations described in Section 2.1 have been developed for squared matrix elements we will analyze the fluctuations of $|A_{jk}|^2$. We expect to find that the unfolded matrix elements $|\tilde{A}_{jk}|^2$ have a Thomas-Porter distribution for which $<|\tilde{A}_{jk}|^2>=1$.

We have therefore devised a predictive method for the matrix elements in the energy representation which is partly exact and partly statistical. This type of information is crucial, for example, when applying the Golden Rule to radiative transitions. In the next chapter we will check our predictions on a numerical model.

3. NUMERICAL MODEL - THE COUPLED ROTATORS

The coupled rotators model (CRM) describes two angular momenta, $\vec{L}$ and $\vec{M}$, coupled to some external field and also among themselves

$$H = L_z + M_z + L_x M_x \tag{3.1}$$

Similar algebraic Hamiltonians were found to give good predictions when applied to nuclei (Williams and Koonin, 1982) and Josephson junctions (DiRienzo and Young, 1983), spin clusters on lattices (Muller, 86) and molecules (Alhassid, 1986). We have extensively studied the CRM both classically and quantum mechanically (Feingold, Moiseyev and Peres, 1984; Feingold and Peres, 1983, 1985). In the following we will give a short review on the behavior of (3.1) and then we will numerically check our predictions of Chapter 2.

The classical equations of motion for the CRM are obtained from the Poisson brackets formalism, for example

$$\dot{L}_x = [L_x, H]_{PB} = -L_y \tag{3.2}$$

where we have used

$$\begin{aligned} [L_m, L_n]_{PB} &= \sum \varepsilon_{mns} L_s \\ [M_m, M_n]_{PB} &= \sum \varepsilon_{mns} M_s \\ [L_m, M_n]_{PB} &= 0 \end{aligned} \tag{3.3}$$

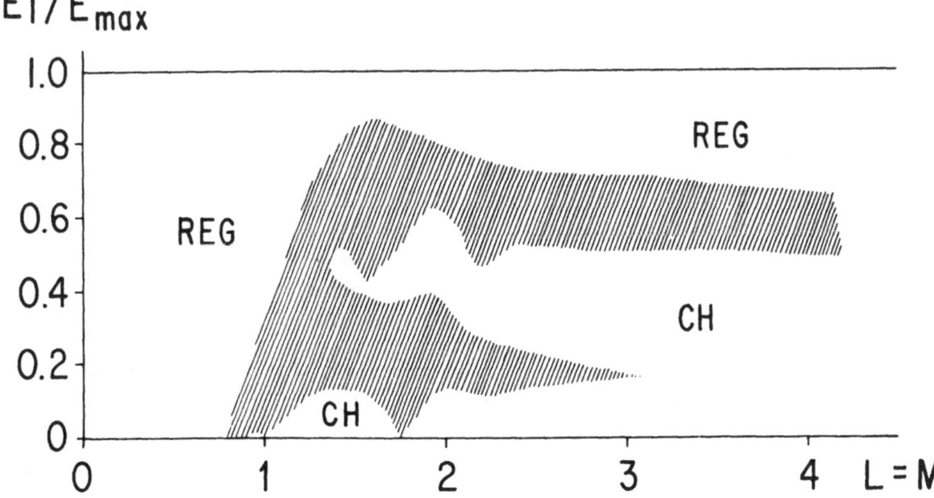

Fig. 4. The regular and chaotic domains of the Hamiltonian (3.1) when the constants of motion L^2 and M^2 are equal. In the hatched region we find both regular and chaotic orbits. The figure is symmetric with respect to the E=0 axis.

There are three constants of motion: E, L and M where $L^2=\sum_{i=1}^{3} L_i^2$. In Fig. 4, a planar section of the $3D$ parameter space, $L=M$, is shown; we will do all our quantum mechanical calculations at $L=M=3.5$. Here, as we change the energy, all three possible behaviors are observed: regular, chaotic and mixed. Specifically, while for $|E|<6.6$ the behavior is chaotic, for $|E|>9.1$ it is regular. Otherwise, both regular and chaotic trajectories are found at the same energy. Notice that while L can grow indefinitely, the energy is bounded, $|E|<E_{max}$, where

$$E_{max} = \begin{cases} 2L & \text{if} \quad L<1 \\ 1+L^2 & \text{if} \quad L>1 \end{cases} \tag{3.4}$$

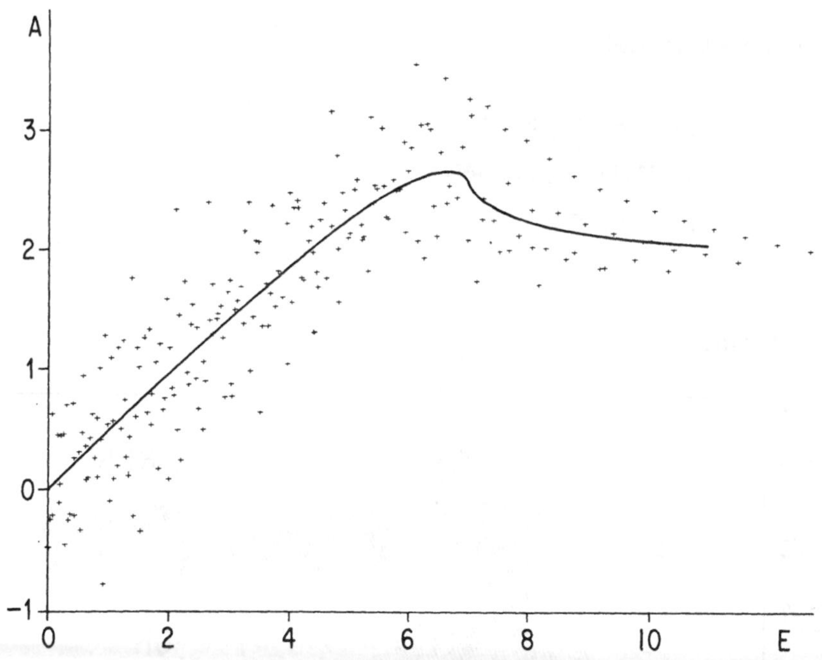

Fig. 5. The diagonal elements of A are represented by the small crosses. For every energy there corresponds only one cross. The continuous line is the microcanonical average of A as a function of energy. Notice that the chaotic region ends at $E\simeq6.6$ and there is no obvious reason our predictions should work much above that.

The most important advantage of CRM over canonical Hamiltonians (like the Henon-Heiles model for example) shows up in the quantization. Here the Hamiltonian matrix is finite and no truncation of the basis set is needed. Using $L^2=\hbar^2 l(l+1)$ and $M^2=\hbar^2 m(m+1)$, we keep $L=M=3.5$ constant and adjust the value of $\hbar$ by changing the l and m quantum numbers. The dimension of the Hamiltonian matrix is $(2l+1)(2m+1)$ such that it diverges as $\hbar \to 0$. In our calculations $l=m=20$ and therefore $\hbar=0.17078....$

In order to check our predictions regarding matrix elements we have chosen $A=L_z+M_z$ as our operator. In Fig. 5 we check the equality among the diagonal elements of A in the energy representation and its microcanonical average for various energies (see Eq. (2.3)). It can be seen that the exact quantum results

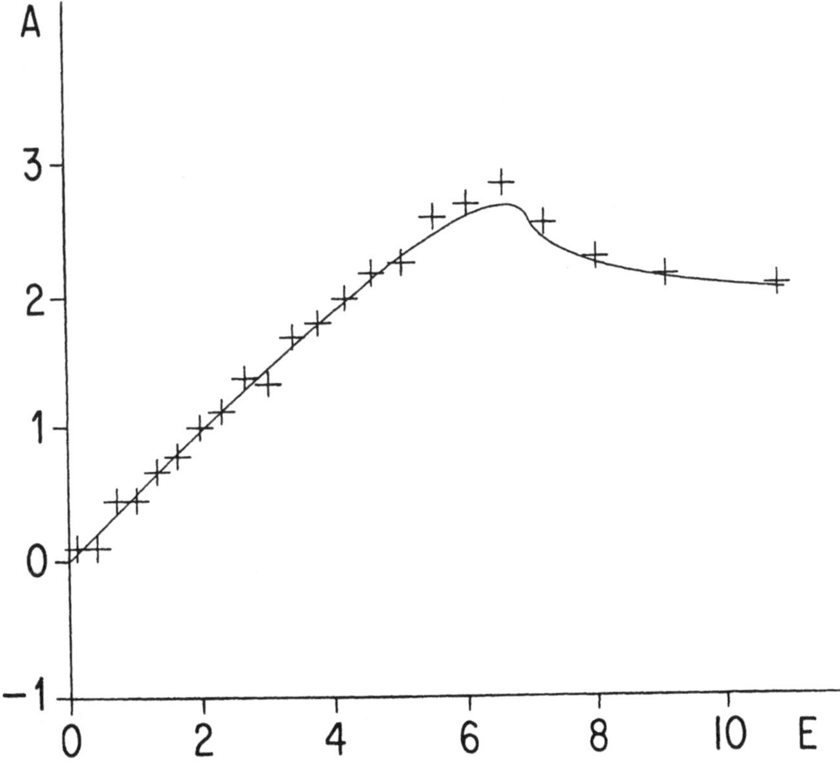

Fig. 6. Same as in Fig. 5, only here the big crosses denote averages over 22 groups of ten consecutive small crosses each.

have large fluctuations around the semiclassical estimate. What is the relation between these two entities? As emphasized in Chapter 2, the fluctuations are a result of the finiteness of $\hbar$. The exact diagonal elements are random variables with average given by the semiclassical estimate and variance which goes to zero whenever $\hbar \to 0$. In order to confirm the first part of this claim we have averaged groups of ten small crosses from Fig. 5 and compared them to the microcanonical average in Fig. 6. Here the agreement between the two was found to be very good. Regarding the variance of the distribution on the other hand, we have argued (Feingold and Peres, 1986) that it is twice the variance of the off-diagonal elements which are close to the diagonal. This prediction was found to be in good agreement with numerical simulation for the CRM. Notice that the relation between diagonal and off-diagonal variances is the same as the one obtained (for H_{ij}) in the frame of random matrix theory (result II.).

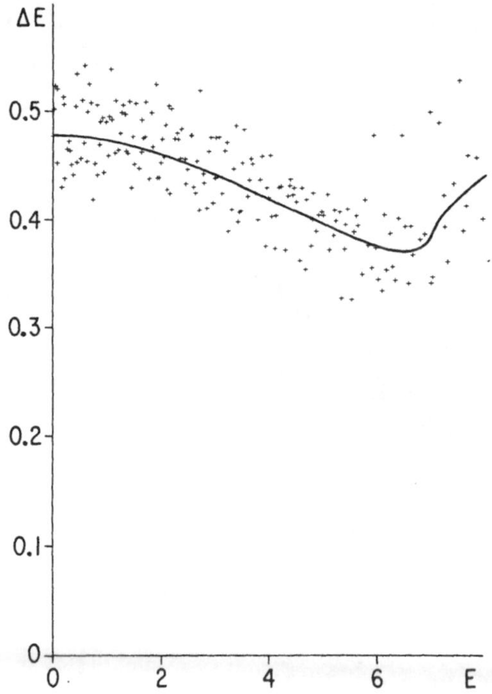

Fig. 7. The two sides of Eq. (2.6) are compared. The small crosses denote the exact quantum energy width of the A matrix as defined by (2.5), for various energies (every row of the matrix gives some value for the ΔE and corresponds to some energy). The continuous line is obtained from the last expression in (2.6); it represents the semiclassical estimation for ΔE.

Now we go over to the issue of the off-diagonal matrix elements. The first method devised in Chapter 2 for estimating matrix elements (i.e. the method of averages) is verified in Fig. 7 for $n=1$ (see Eqs. (2.4), (2.5) and (2.6)). Here we can see find a similar type of agreement with that found in Fig. 5. In this case also, when the quantum behavior is smoothed by group averaging very good agreement with the semiclassical estimate is obtained.

The first impression might be that the second method leading to the off-diagonal matrix elements (see Eq. (2.11)) is much more direct and therefore the method of averages is not at all useful. However, calculating the power spectrum for chaotic trajectories turns out to be difficult and unprecise. Thus, the price we are paying for directness is accuracy. That is because $A(t)$ is non-periodic and moreover it does not decay at infinity. Using a wisdom from the theory of stochastic processes (Born and Wolf, 1980; Davenport and Root, 1958) we estimate our power spectrum by the periodogram

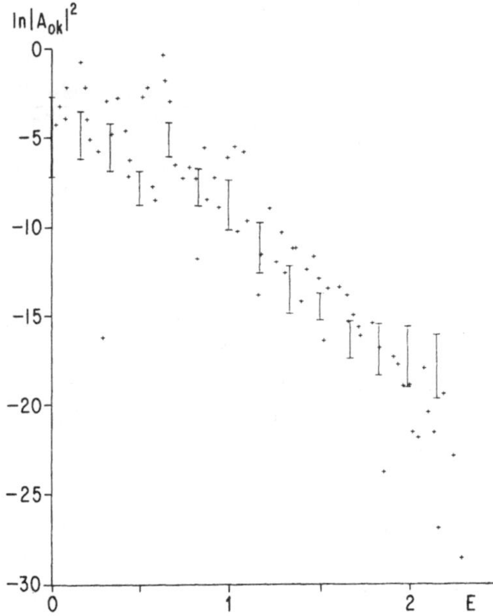

Fig. 8. The off-diagonal matrix elements in the $E=0$ row. The small crosses are the exact quantum mechanical results. On the other hand, the error bars represent the right hand side of (2.11). The power spectrum is averaged over an ensemble of 30 trajectories, each one iterated for T=6694 (about 1000 oscillations of the CRM).

$$S(\omega,T) = \frac{1}{T} | \int_0^T A(t) e^{i\omega t}\, dt |^2 \tag{3.5}$$

Although $\lim_{T\to\infty} S(\omega,T)$ is not numerically computable, for an ergodic system, taking $T\to\infty$ is equivalent to averaging the finite T power spectrum in (3.5) over an infinite ensemble of initial conditions. In Fig. 8 the exact quantum off-diagonal matrix elements (small crosses) are compared with the averaged $S(\omega,T)$ (error bars). The errors are a result of the finiteness of the sample and are calculated in the standard way. Considering that the comparison is over ≈10 orders of magnitude, the agreement is very good. For small values of $|A_{jk}|^2_{SC}$ ($\approx 10^{-12}$) some disagreement emerges from the round off numerical noise.

This ends up our semiclassical predictions. Because of the big error bars in Fig. 8 and of the fact that the method of averages becomes increasingly difficult as n increases, we will use an empirical $|A_{jk}|^2_{SC}$, $|A_{jk}|^2_{EM}$, rather then the semiclassically predicted one. $|A_{jk}|^2_{EM}$ is obtained by a Gaussian smoothing of the exact quantum $|A_{jk}|^2$

$$|A_{jk}|^2_{EM} = \frac{\sum_{k',j'} |A_{j'k'}|^2 e^{-\frac{(\tilde{E}_{j'}-\tilde{E}_j)^2}{2\sigma^2}} e^{-\frac{(\tilde{E}_{k'}-\tilde{E}_k)^2}{2\sigma^2}}}{\sum_{k',j'} e^{-\frac{(\tilde{E}_{j'}-\tilde{E}_j)^2}{2\sigma^2}} e^{-\frac{(\tilde{E}_{k'}-\tilde{E}_k)^2}{2\sigma^2}}} \tag{3.6}$$

This empirical procedure, originally used by Alhassid and Levine (Alhassid and Levine, 1986) can be in principle avoided by extending our semiclassical calculations to a more accurate estimation of $|A_{jk}|^2_{SC}$. The value of σ in (3.6) has to be chosen as small as possible such that the decay of $|A_{jk}|^2_{EM}$ is mainly monotonic as we increase the energy distance from the diagonal.

In order to numerically analyze the fluctuations, we unfold the matrix elements in a 40×40 block of the 441×441 A matrix using (2.7), with $|A_{jk}|^2_{EM}$ replacing $|A_{jk}|^2_{SC}$. The block is defined by $0<E_j,E_k<1.24$ such that its diagonal overlaps with the matrix diagonal. This energy range is deeply embedded in the classically chaotic region. In Fig. 9 we compare the distribution of $|\tilde{A}_{jk}|^2$ with the Thomas-Porter distribution and find excellent agreement. As in Fig. 8, here the size of the matrix element is plotted on a logarithmic scale such that the two curves agree over about six orders of magnitude. We should mention that the normalization procedure using $|A_{jk}|^2_{EM}$ instead of $|A_{jk}|^2_{SC}$ is not very accurate. For example, values of $<|\tilde{A}_{jk}|^2>$ as low as 0.8 have been obtained when applying (3.6) (compare to the theoretical value which is 1). Therefore we have decided to take $<|\tilde{A}_{jk}|^2>$ as a free parameter in our computations.

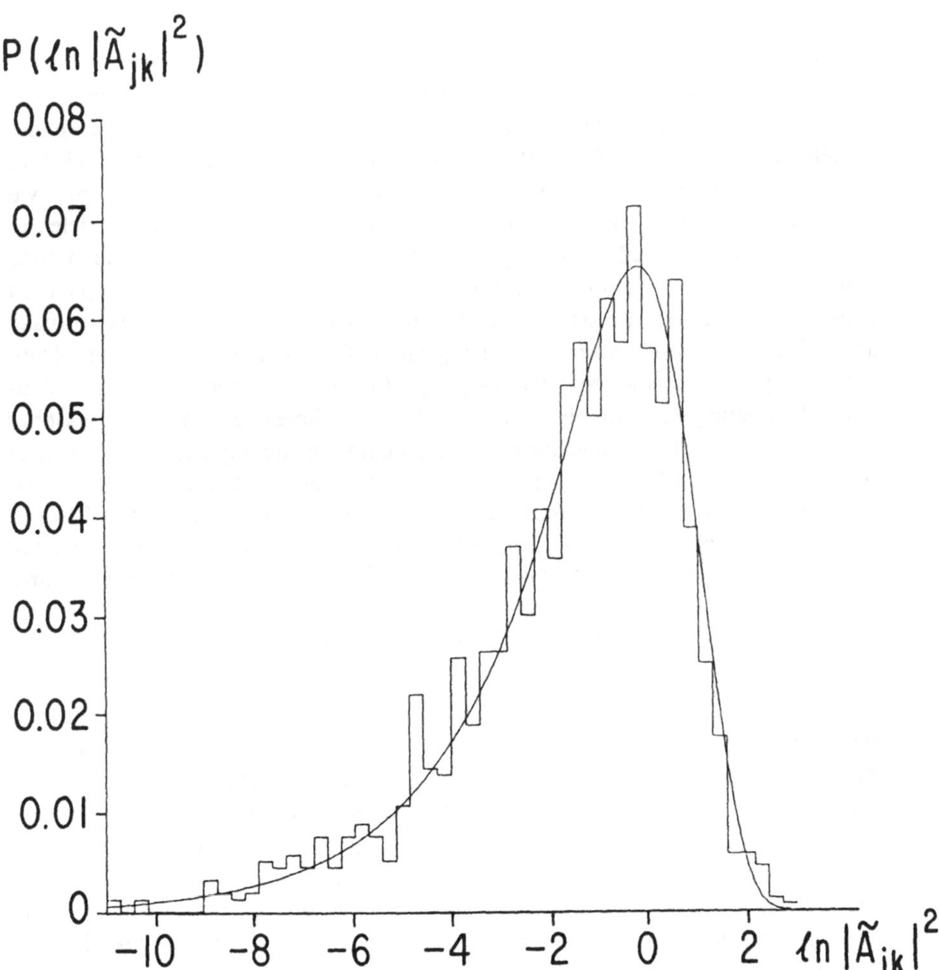

Fig. 9. The distribution of $|\tilde{A}_{jk}|^2$ on a logarithmic scale. The histogram is numerically calculated for all the 1600 matrix elements in the 40×40 submatrix $0<E_j,E_k<1.24$. The continuous line represents the Thomas-Porter distribution (see Eq. (2.16)).

4. OPEN PROBLEMS

In this section we will sketch some of the unsettled questions. To few of them we already have tentative answers which will also be shortly reviewed below.

The main question concerns the relevance of our results to experiments; can the ergodic quantization for matrix elements become into a computational tool for real systems like molecules, atoms or nuclei? Suppose we can write the Hamiltonian explicitly and that the classical motion is chaotic enough such that the ergodic assumption would work. In order to calculate the power spectrum we would have to iterate the several hundreds (six times the number of particles in the system) of first order ordinary differential equations for long enough time and enough initial conditions to get a fair degree of accuracy. This computation does not have a stepwise nature and therefore there is no obvious way to make approximations. On the other hand, using the method of averages, one can obtain more exact information about the $|A_{jk}|^2_{SC}$ function for every new M_{2n} calculated. Estimating M_{2n} implies performing $2N-1$ dimensional integrations over energy surfaces. The complexity of the integrand will also quickly increase with n. However, methods for estimating multidimensional integrals are well developed in the context of cross section calculations in particle physics. Thus, it seems reasonable to expect that, provided some appropriate approximation methods will be found, the method of averages could be applied to real systems. Once $|A_{jk}|^2_{SC}$ is found, we can predict the behavior of the random fluctuations, as described in Chapters 2 and 3, without any additional information.

The next problem deals with the non-ergodic regime. We know how to work out semiclassical mechanics for integrable systems and now we learned how to estimate matrix elements for strongly chaotic systems. Could we extend our arguments to the large region in parameter space where classically both regular and chaotic trajectories coexist? We shall start with an observation regarding the semiclassical part of the ergodic quantization. As we can easily see in Fig. 5, the agreement between the semiclassical estimation and the exact quantum result does not end at the limit of the chaotic region but continues all the way into the mixed and regular regions. This feature is even more strikingly emphasized in Fig. 6. A possible explanation for the agreement in non-ergodic regime is that the small crosses in Fig. 5 obtain their value from different regular regions for consecutive energies; those regular regions are spread all over the energy surface. When group averaging over ten regular regions we roughly get a representative from every part of the energy surface and therefore we practically perform a microcanonical average. A similar behavior is observed for the energy width, ΔE (see Eq. (2.5) and (2.6)). It is not yet clear if this extended agreement is a general feature or just a particular property of the CRM. What can be said about fluctuations of matrix elements in the mixed and regular regimes? Those are clearly not randomly distributed (see Fig. 5 for E>7). Actually, diagonal elements in the regular regime have been shown to have a lattice structure with a lattice constant which scales like $\hbar$ (Peres, 1984). There is no such prediction for the off-diagonal case. We therefore employed the same method as in the chaotic regime in order to study the distribution of fluctuations. It turns out that as we approach the regular regime, their distribution (on the logarithmic scale, see Fig. 9) strongly widens. This widening is partly a result of very small matrix elements. They correspond to broken selection rules of the completely integrable system. The widening of the distribution of fluctuations also points out the existence of richer quantum structure in the regular than in the chaotic case. However, we have no quantitative prediction yet for the non-ergodic regime regarding fluctuations of the matrix elements.

The third and last problem is whether the type of arguments presented in this paper could be applied to other quantum objects as well. Berry has shown using semiclassical arguments that the chaotic wave functions are Gaussian random functions (Berry, 1977)

$$P\,[\psi(\vec{r})] = \frac{1}{\sqrt{2\pi\,\langle\psi^2(\vec{r})\rangle}}\; e^{-\frac{\psi^2}{2\langle\psi^2(\vec{r})\rangle}} \tag{4.1}$$

In order to semiclassically estimate the variance of (4.1), the assumption that the Wigner function is homogeneous over the energy surface is used

$$W(\vec{r},\vec{p}) \approx \frac{\delta[E-H(\vec{r},\vec{p})]}{\iint d\vec{r}d\vec{p}\;\delta[E-H(\vec{r},\vec{p})]} \tag{4.2}$$

Therefore we obtain

$$\langle\psi^2(\vec{r})\rangle = \int d\vec{p}\;W(\vec{r},\vec{p}) = \frac{\int d\vec{p}\;\delta[E-H(\vec{r},\vec{p}]}{\iint d\vec{r}d\vec{p}\;\delta[E-H(\vec{r},\vec{p})]} \tag{4.3}$$

This prediction was found to be in good agreement with numerical data for the stadium model (Shapiro and Goelman, 1984). Moreover, (4.1) is also predicted by random matrix theory (see also Eq. (2.15)). The algorithm used for analyzing the wave functions might look different at first from that of ergodic quantization. However, the difference results from investigating an object which changes sign and should disappear if studying $|\psi(\vec{r})|^2$.

We conclude that ergodic quantization supplies a combination of exact and statistical information (using a unique algorithm) for classically ergodic systems regarding all three ingredients of the Schrodinger equation: spectrum, matrix elements (of perturbations) and wave functions. Therefore it qualifies as an analog of the Einstein-Brillouin-Keller semiclassical quantization method in the case of strongly chaotic systems.

AKNOWLEDGMENTS

Quite a lot of the ideas reviewed in this paper are a result of discussions and collaboration with A. Peres and Y. Alhassid. I also want to thank M. Berry, S. Fishman, L. Kadanoff, I. Percival, S.A. Rice and N. Soker for useful comments and suggestions. Support from a Dr. Chaim Weizmann post-doctoral fellowship is gratefully acknowledged.

REFERENCES

Alhassid, Y. 1986, Phys. Rev. Lett. 57 539-542

Alhassid, Y. and Levine, R.D. 1986, To appear in Phys. Rev. Lett.

Berry, M.V. 1977, J. Phys. A 10 2083-2091

Berry, M.V. 1981, *Chaotic Behavior of Deterministic Systems*, Proceedings of the Les Houches Summer School, Eds. R.H.G. Helleman and G. Ioos, North Holland, Amsterdam

Berry, M.V. 1985, Proc. Roy. Soc. London A 400 229-251

Berry, M.V. and Tabor, M. 1977, Proc. Roy. Soc. London A 356 375-394

Bohigas, O., Giannoni, M.J. and Schmit, C. 1984, Phys. Rev. Lett. 52 1-4

Born, M. and Wolf, E. 1980, *Principles of Optics*, Pergamon, Oxford

Brody, T.A., Flores, J., French, J.B., Mello, P.A., Pandey, A. and Wong, S.S.H. 1981, Rev. Mod. Phys. 53 385-479

Casati, G., Valz-Gris, F. and Guarneri, I. 1980, Nuovo Cimento Lett. 28 279-282

Davenport, W.B. Jr and Root, W.L. 1958, *An Introduction to the Theory of Random Signals*, McGraw-Hill, New York

DiRienzo, A.L. and Young, R.A. 1983, Am. J. Phys. 51 587-597

Dyson, F.J. and Mehta, M.L. 1963, J. Math. Phys. 4 701-719

Feingold, M., Moiseyev, N. and Peres, A. 1984, Phys. Rev. A 30 509-511

Feingold, M., Moiseyev, N. and Peres, A. 1985, Chem. Phys. Lett. 117 344-346

Feingold, M. and Peres, A. 1983, Physica D 9 433-438

Feingold, M. and Peres, A. 1984, Phys. Rev. A 31 2472-2476

Feingold, M. and Peres, A. 1986, Phys. Rev. A 34 591-595

Gelbart, W.M., Rice, S.A. and Freed, K.F. 1972, J. Chem. Phys. 57 4699-4712

Gaudin, M. 1961, Nucl. Phys. 25 447-458

Gutzwiller, M.C. 1967, J. Math. Phys. 8 1979-2000

Gutzwiller, M.C. 1969, J. Math. Phys. 10 1001-1020

Gutzwiller, M.C. 1970, J. Math. Phys. 11 1791-1806

Gutzwiller, M.C. 1971, J. Math. Phys. 12 343-358

Gutzwiller, M.C. 1980, Phys. Rev. Lett. 45 150-153

Haller, E., Koppel, H. and Cederbaum, L.S. 1984, Phys. Rev. Lett. 52 1665-1668

Heller, E.J. and Sundberg, R.L. 1984, in *Chaotic Behavior in Quantum Systems*, Proceedings of the Como Conference on Quantum Chaos, Ed. G. Casati, Plenum Press, New York 255-292

Keller, J.B. 1958, Ann. Phys. N.Y. 4 180-188

Khintchine, A. 1934, Math. Ann. 109 604-617

Lichtenberg A.J. and Lieberman M.A. 1983, *Regular and Stochastic Motion*, Springer Verlag, Berlin

McDonald, S.W. and Kaufman, A.N. 1979, Phys. Rev. Lett. 42 1189-1191

Mehta, M.L. 1960, Nucl. Phys. 18 395-419

Mehta, M.L. 1967, *Random Matrices*, Academic Press, New York

Muller, G.H. 1986, Phys. Rev. A 34 3345-3355

Pechukas, P. 1983, Phys. Rev. Lett. 51 943-946

Peres, A. 1984, Phys. Rev. Lett. 53 1711-1713

Porter, C.E. and Thomas, R.G. 1956, Phys. Rev. 104 483-491

Seligman, T.H., Verbaarschot, J.J.M. and Zirnbauer, M.R. 1984, Phys. Rev. Lett. 53 215-217

Shapiro, M. and Goelman, G. 1984, Phys. Rev. Lett. 53 1714-1717

van Vleck, J.H. 1928, Proc. Natl. Acad. Sci. U.S.A. 14 178-188

Wiener, N. 1930, Acta. Math. 55 117-130

Williams, R.D. and Koonin, S.E. 1982, Nucl. Phys. A 391 72-81

CANTORI AND QUANTUM MECHANICS

T GEISEL, G RADONS AND J RUBNER

ABSTRACT

This paper investigates the role of cantori for the quantum dynamics of a chaotic system with two degrees of freedom. They can act as barriers more drastically than in classical mechanics and inhibit the diffusive growth of mean square displacements. The asymptotic distribution decays exponentially in their vicinity. The penetration probability and depth of a KAM-torus are also determined.

The advances made in the understanding of chaos in classical Hamiltonian systems[1] have stimulated the search for analogous phenomena in quantum mechanics. It has turned out that the two theories may exhibit strongly contrasting properties. E.g. classical systems may exhibit large-scale chaotic motions in the form of diffusion, which are suppressed in quantum mechanics[2-10]. Especially in the limit $t\to\infty$ quantum mechanics does not show certain phenomena that exist in its classical limit.

This paper deals with fractals that play a role as partially penetrable barriers in classical mechanics. In systems with two degrees of freedom, invariant KAM-tori confine the chaotic motions to certain regions of phase space[1]. With increasing nonlinearity these tori can turn into invariant cantor sets, so-called cantori, and become partially penetrable.[11,12] We have asked whether the classical KAM-tori and cantori affect the dynamics of a quantum system that is classically chaotic. It turned out that they do play a role as dynamical barriers. There is only a small transition probability into states that

* Work supported by Deutsche Forschungsgemeinschaft
** Heisenberg fellow

are classically inaccessible, which scales as $\hbar^{2.5}$ for a KAM-torus. The asymptotic probability distribution decays exponentially near KAM-tori and cantori; the penetration depth scales as $\hbar^{0.66}$. One may interpret this as a KAM-localization mechanism in quantum systems; its origins are the classical KAM-tori and cantori, as opposed to quantum interferences (Anderson localization) discussed previously.[5,6,8,9] Amazingly, (classical) cantori appear to act as barriers even more drastically than in the classical case: e.g. they can entirely inhibit the diffusive growth of mean-square displacements, whereas classically their presence only slows down the growth.[11,12] This means that a cantorus virtually prevents the quantum system from exploring regions of phase space that are eventually explored by its classical analog. These results were obtained for the model of a kicked quantum rotator, where a single confining KAM-torus can be isolated. We expect, however that they are generally characteristic of the permeability of KAM-tori and cantori and not of a particular model. Thus they may also be relevant e.g. in laser chemistry for multiphoton dissociation of molecules, where the regions of phase space relating to bound and dissociated states can be separated by KAM-tori or cantori. Their permeability as studied here then determines the dissociation rate. As a matter of fact, in a recent numerical study of multiphoton dissociation, Brown and Wyatt[13] have observed that dissociation can indeed be inhibited considerably by the presence of cantori.

An appropriate model for the investigation of the above problems is the kicked planar rotator,[2-6,10,14] defined by the Hamiltonian

$$\hat{H} = \frac{\hat{p}^2}{2I} + k \cos \hat{\theta} \sum_{n=-\infty}^{\infty} \delta(t-nT) \qquad (1)$$

where $\hat{\theta}$ and $\hat{p}$ are angle and angular momentum operators, I is the moment of inertia, and the kicking strength k determines the degree of nonlinearity. Classically the equation of motion following from Eq. (1) is Chirikov's standard map[15] (assuming $I = T = 1$)

$$p_{t+1} = p_t + k \sin \theta_{t+1} \tag{2}$$

$$\theta_{t+1} = \theta_t + p_t \tag{3}$$

where p_t and θ_t denote the coordinates immediately after the kick at time t. Its classical phase space (Fig. 1) according to the Poincaré-Birkhoff theorem and KAM-theorem[1] typically exhibits elliptic and hyperbolic fixed points and KAM-tori (here as circles). The KAM-tori are invariant sets and thus cannot be penetrated by the chaotic orbits that develop near hyperbolic fixed points.

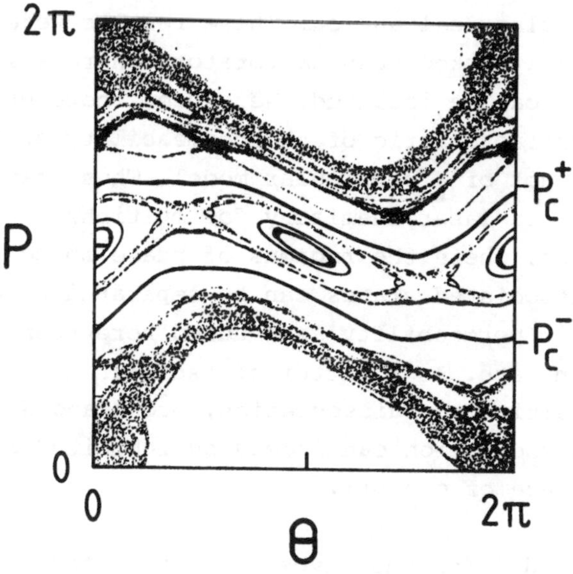

Fig. 1: Classical phase space for $k = k_c = 0.9716$ showing 5000 iterates of 13 initial points (θ_0, p_0) according to Eqs. (2,3). Orbits starting between the two KAM-tori remain confined to the momentum interval (p_c^-, p_c^+).

With increasing nonlinearity k, the KAM-tori break up into cantori[11,12] (invariant cantor sets) and become partially penetrable. There is a critical nonlinearity $k = k_c = 0.9716$ with universal scaling properties,[16] where the last KAM-tori divide the phase space into cells along the p-axis. This is the situation shown in Fig. 1. It follows that the asymptotic distribution

$$\rho(\theta,p) = \lim_{N\to\infty} \frac{1}{N} \sum_{t=0}^{N-1} \rho_t(\theta,p) \tag{4}$$

remains confined to a momentum interval $(p_c^-,p_c^+) = (2.015, 4.269)$, e.g. if an initial distribution $\rho_0(\theta,p) = (1/2\pi)\delta(p-p_0)$ with $p_0 = 3.2$ is chosen. Above k_c these KAM-tori are broken, giving rise to a diffusive motion along the p-axis[15] similar to the one in dissipative systems.[17]

The quantum dynamics can be expressed in terms of the time evolution operator $\hat{U}$

$$|\psi_{t+1}\rangle = \hat{U}|\psi_t\rangle \tag{5}$$

where $|\psi_t\rangle$ denotes the quantum state immediately after the kick at time t, and for Eq. (1) $\hat{U}$ follows as

$$\hat{U} = \exp(-\frac{i}{\hbar}\, k \cos\hat{\theta})\, \exp(-\frac{iT}{\hbar}\frac{\hat{p}^2}{2I}) \quad . \tag{6}$$

With $\hat{p} = -i\hbar\partial/\partial\theta$ it is clear that quantum effects depend on the ratio $\hbar T/I$. In practice the classical limit can be approached by varying I or T. Equivalently we may consider $\hbar T/I$ as an effective Planck's constant that can be varied. Without loss of generality we will henceforth assume I = T = 1 and vary $\hbar$ instead (i.e. measure $\hbar$ in units of I/T). Iteration of the time evolution operator yields

$$|\psi_t\rangle = \hat{U}^t|\psi_0\rangle \quad . \tag{7}$$

As an initial state we use a momentum eigenstate $|\psi_0\rangle = |p_0\rangle$ with $\hat{p}|p_0\rangle = p_0|p_0\rangle$ and $p_0 = 3.2$. This is the analog of the initial distribution $\rho_0(\theta,p)$ discussed in the classical case above. We define the asymptotic distribution of angular momenta as a time average

$$P(p|p_0) = \lim_{N\to\infty} \frac{1}{N} \sum_{t=0}^{N-1} |\langle p|\psi_t\rangle|^2 \quad , \tag{8}$$

its existence and convergence was verified numerically by varying N. With Eq. (7) P can be written as

$$P(p|p_0) = \lim_{N\to\infty} \frac{1}{N} \sum_{t=0}^{N-1} |\langle p|\hat{U}^t|p_0\rangle|^2 \ . \tag{9}$$

The operator $\hat{U}$ can in principle be represented in the momentum basis and iterated. Analytically $\langle p|\hat{U}|p'\rangle$ can be expressed in terms of Bessel functions ($p = m\hbar$)

$$\langle m\hbar|\hat{U}|m'\hbar\rangle = \frac{i^{m-m'}}{\hbar} e^{-i\hbar m^2/2} J_{m'-m}(k/\hbar) \ . \tag{10}$$

For numerical purposes, however, it is more convenient to iterate $\hat{U}$ using forward and backward Fast-Fourier-Transform.[5]

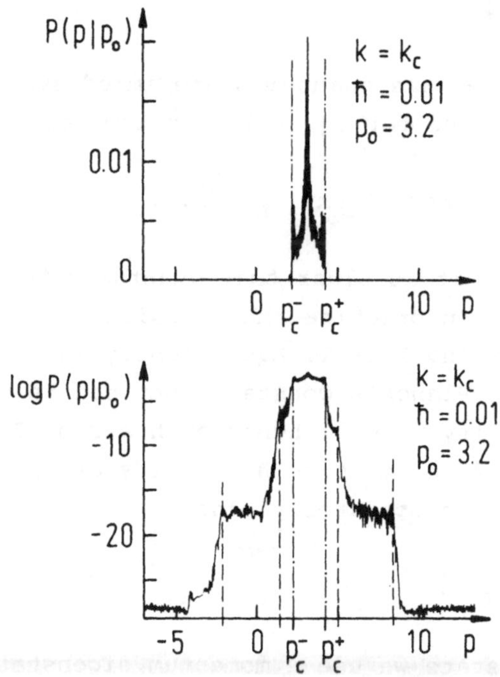

Fig. 2: Asymptotic momentum distribution Eq. (8) a) on a linear, b) on a logarithmic scale. Borders of classical confinement due to KAM-tori (dot-dashed lines) and cantori (dashed lines) are indicated. The initial momentum p_0 was in the interval (p_c^-, p_c^+).

We have calculated the asymptotic momentum distribution $P(p|p_0)$ for various values of k and $\hbar$.[18] This quantity is the analog of the classical distribution $\bar{\rho}(\theta,p)$ (Eq. (4)) projected onto the momentum axis. Fig. 2a shows for $k = k_c$ that most of the time averaged probability remains confined to the classically accessible interval (p_c^-, p_c^+). This demonstrates the role of KAM-tori as barriers in quantum systems. The logarithmic display in Fig. 2b shows in more detail how the probability leaks through the torus. It decays exponentially into the classically inaccessible region

$$P(p|p_0) \propto e^{-\lambda|p-p_e|} \tag{11}$$

where $p_e = p_c^{\pm}$. Moreover the figure shows that there is also an exponential decay near cantori as indicated by the dashed lines. The same is observed for $k = 1.1 > k_c$ where the tori near $p_c^{\pm}$ have turned into cantori.[18] We can understand the exponential decays in terms of Wigner functions corresponding to quasienergy states $|j\rangle$ (eigenfunctions of $\hat{U}$). $P(p|p_0)$ can be expressed as

$$P(p|p_0) = \sum_j |\langle p|j\rangle|^2 \cdot |\langle p_0|j\rangle|^2 \tag{12}$$

where $|\langle p|j\rangle|^2$ is the projection of the Wigner function $W_j(\theta,p)$ belonging to $|j\rangle$. It is known that Wigner functions that are located on tori in the semiclassical limit decay exponentially.[19] From this simple argument one would expect that $\lambda \sim \hbar^{-1}$ for $\hbar \to 0$. The same behaviour would be found for penetration of a potential step due to tunneling. What we observe, however, is an algebraic dependence of λ as $\hbar^{-0.66}$ (Fig. 3); we attribute it to the complexity of phase space near the KAM-torus. Penetration of a torus is obviously more complicated than tunneling through a potential barrier. The exponent -0.66 might be related to critical exponents at the stochastic transition, where in the semiclassical limit $\hbar$ acts as a relevant variable.[20]

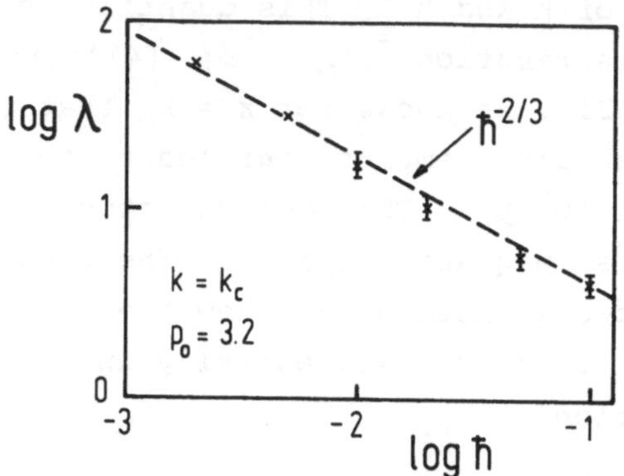

Fig. 3: Inverse penetration depth λ for the KAM-torus ($k = k_c$).

The quantum permeability of a KAM-torus can be characterized best by the asymptotic transition probability $\bar{W}$ into classically inaccessible regions, i.e. the total probability outside the dot-dashed lines in Fig. 2

$$\bar{W} = \sum_{p<p_c^-} P(p|p_0) + \sum_{p>p_c^+} P(p|p_0) \qquad (13)$$

Fig. 4a shows that this transition probability decreases considerably when k_c is approached from above, i.e. when a KAM-torus builds up. The suppression of $\bar{W}$ between k_c and 1.1 shows that cantori can act as barriers as well. At the critical nonlinearity k_c (Fig. 4b) the permeability due to quantization exhibits a power law dependence $\bar{W} \sim \hbar^{2.5}$. If we choose p_0 in the much larger cells around integer resonances, we find that the exponent is changed. This appears to be due to partial localization before the wave function reaches the torus.

We quote some results on time dependent quantities, which will be reported in more detail elsewhere.[18] Where tori and cantori act as barriers, the transition probability increases algebraically in time before turning into quasiperiodic behaviour. The same is true for the mean-square displacement of momentum,

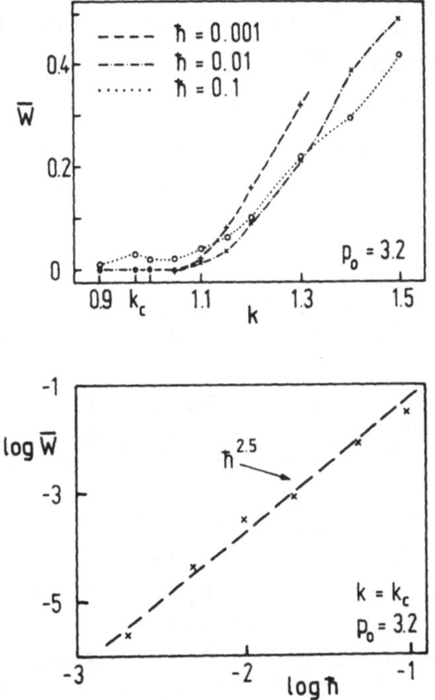

<u>Fig. 4:</u> Asymptotic transition probability $\bar{W}$ into classically inaccessible regions a) as the torus breaks up above k_c, b) due to quantum effects for $k = k_c$.

but only for larger k (k > 1.2). For smaller k ($k_c < k \leq 1.2$, $10^{-4}<\hbar<10^{-1}$) it remains bounded within the cantori ($<\Delta p_t^2><10^{-1}$) and does not show any diffusive growth. This should be contrasted with the classical case, where it grows to infinity.[15] The role of classical cantori as barriers in quantum systems thus is more drastical than even in classical mechanics. This property represents a striking difference in the performance of classical chaotic systems and their quantum analogs and can be used as a test in experiments on quantum chaos. Classically probability can flow constantly across a cantorus such that after a sufficiently long time the system has essentially penetrated it. Most of the quantum mechanical probability, however, remains confined asymptotically (see also Fig. 4a for $k_c < k < 1.1$). The classical and the quantum system can be found in different states after a sufficiently long time. This phenomenon should be observable in experiments where a cantorus separates two qualitatively different states, as it is the case for multiphoton dissociation of diatomic molecules.[13] It might even explain why diatomic molecules show little dissociation in comparison e.g. to SiF_4,[21] where cantori do not act as barriers due to a higher dimensional phase space.

References

1. For a review see e.g. M.V. Berry, in Topics in Nonlinear Dynamics, ed. S. Jorna, AIP Conf. Proc. Vol. 46 (New York 1978) p. 16.
2. G. Casati, B.V. Chirikov, F.M. Izrailev, and J. Ford, in Stochastic Behaviour in Classical and Quantum Hamiltonian Systems, ed. G. Casati and J. Ford, Lect. Notes in Physics Vol. 93 (Springer, Berlin 1979), p. 334.
3. B.V. Chirikov, F.M. Izrailev, and D.L. Shepelyanski, Sov. Sci. Rev. Sect. C 2, 209 (1981).
4. T. Hogg and B.A. Huberman, Phys. Rev. Lett. 48, 711 (1982) and Phys. Rev. A 28, 22 (1983).
5. S. Fishman, D.R. Grempel, and R.E. Prange, Phys. Rev. Lett. 49, 509 (1982); D.R. Grempel, R.E. Prange, and S. Fishman, Phys. Rev. A 29, 1639 (1984).
6. J.D. Hanson, E. Ott, and M. Antonsen, Phys. Rev. A 29, 819 (1984).
7. R.V. Jensen, Phys. Rev. Lett. 49, 1365 (1982).
8. R. Blümel and U. Smilansky, Phys. Rev. Lett. 52, 137 (1984).
9. G. Casati, B.V. Chirikov, and D.L. Shepelyanski, Phys. Rev. Lett. 53, 2525 (1984).
10. D.L. Shepelyanski, Physica 8 D, 208 (1983).
11. R.S. MacKay, J.D. Meiss, and I.C. Percival, Physica 13 D, 55 (1984).
12. D. Bensimon, and L.P. Kadanoff, Physica 13 D, 82 (1984).
13. R.C. Brown and R.E. Wyatt, Phys. Rev. Lett. 57, 1 (1986).
14. M.V. Berry, N.L. Balazs, M. Tabor, and A. Voros, Ann. Phys. (N.Y.) 122, 26 (1979).
15. B.V. Chirikov, Phys. Rep. 52, 623 (1979).
16. J.M. Greene, J. Math. Phys. 20, 1183 (1979), S.J. Shenker and L.P. Kadanoff, J. Stat. Phys. 27, 631 (1982).
17. T. Geisel and J. Nierwetberg, Phys. Rev. Lett. 48, 7 (1982), T. Geisel and S. Thomae, Phys. Rev. Lett. 52, 1936 (1984).
18. T. Geisel, G. Radons, and J. Rubner, to be published.

19. M.V. Berry, Phil. Trans. Roy. Soc. Lond. 287, 237 (1977).

20. D.R. Grempel, S. Fishman, and R.E. Prange, Phys. Rev. Lett. 53, 1212 (1984), and D.R. Grempel, privat comm.

21. N.R. Isenor and M.C. Richardson, Appl. Phys. Lett. 18, 224 (1971).

INFLUENCE OF PHASE NOISE IN CHAOS AND DRIVEN OPTICAL SYSTEMS

L A LUGIATO, M BRAMBILLA, G STRINI AND L M NARDUCCI

1. INTRODUCTION

Noise of both deterministic and stochastic origin has been the object of extensive investigations in recent times. A subject of special interest is the influence of stochastic fluctuations on the dynamics of a system which, under noise-free conditions, displays chaotic behavior. Generally, the aim of this type of investigations is to determine to what extent stochastic noise degrades the characteristic features of deterministic chaos and the temporal structures that emerge along the different routes to chaos. These analyses have led to the development of diagnostic techniques for the identification of chaos even in the presence of a certain amount of stochastic fluctuations.

A very common type of noise results from fluctuations in the parameters that control the interaction of the system of interest with the external environment. The effect of parameter noise on discrete maps has been considered, for example, in Ref. 1. In this article we discuss a test case where the parameter in question is <u>complex</u> (2,3). To be more specific, we consider a class of optical systems driven by an external field. In the slowly varying amplitude approximation, the input field is complex. We focus on the fluctuations of the <u>phase</u> of the slowly varying field, a well known phenomenon that sets a limit to the coherence of laser light, for example, and is responsible for its finite linewidth. The phase diffusion model is the most common tool for the description of this effect (4-6). The main goal of our contribution is to analyze the effect of phase diffusion on selected temporal structures, including deterministic chaos, that emerge as a result of instabilities in the laser with injected signal (LIS) (7) and in optical bistability (OB) (8).

It is intuitively reasonable to expect that the influence of phase noise should become appreciable only when the coherence time of the input radiation becomes comparable to or even shorter than the period of the deterministic oscillations. We find instead that some temporal structures in the LIS are far more sensitive to phase noise than this argument would lead us to believe; by this we mean that the output signals display a strong degradation even when the coherence time of the incident light is much longer than the duration of a single oscillation. In optical bistability, instead the influence of phase fluctuations is less pronounced, at least for the type of pulsing solutions considered in this paper.

An interesting result concerns the appearance of noise-induced oscillations even for parameter values such that the deterministic theory predicts a stable stationary state; this type of modulation is reminiscent of deterministic period-one solutions, apart from the expected irregularities. Thus, in this case, noise does not destroy, but actually aids the development of temporal structures.

Our analysis is interesting not only as an example of complex parametric noise, but should be valuable also in connection with the recent experimental observations of instabilities and self-pulsing in optical bistability (9) and in the laser with an injected signal (10,11). In addition it gives an indication of the degree of frequency stability that must be imposed on the external laser source in order to observe the temporal structures predicted by the deterministic model.

2. THE MODEL

An incident beam with carrier frequency ω_0, is injected into a ring cavity. We call ω_c the cavity resonance that lies nearest to ω_0. The cavity contains a homogeneously broadened two-level medium with transition frequency ω_a. The equations of motion are (8,12)

$$\frac{dx}{d\tau} = -\tilde{\kappa}(i\theta x + x - ye^{i\phi} + 2Cp) \tag{1.1}$$

$$\frac{dp}{d\tau} = xD - (1+i\Delta)p \tag{1.2}$$

$$\frac{dD}{d\tau} = -\tilde{\gamma}\{\tfrac{1}{2}(xp^* + x^*p) + D \pm 1\} \tag{1.3}$$

where ϕ is the phase of the incident field and y is the amplitude, normalized to the square root of the saturation intensity; x is the complex normalized amplitude of the output field; p and D are the scaled atomic polarization and population difference, respectively. The time $\tau = \gamma_\perp t$ is measured in units of the inverse linewidth, $\tilde{\kappa} = \kappa/\gamma_\perp$ is the scaled cavity linewidth, and $\tilde{\gamma} = \gamma_\parallel/\gamma_\perp$ is the ratio between the longitudinal and transverse relaxation rates; $\Delta = (\omega_a - \omega_o)/\gamma_\perp$ is the atomic detuning parameter and $\theta = (\omega_c - \omega_o)/\kappa$ is the cavity mistuning parameter. In Eq. (1.3) the plus sign must be selected for the LIS, for which the population of the two-level medium is inverted and C plays the role of the pump parameter (12). In the case of OB we select the minus sign, and C is interpreted as the usual bistability parameter (8).

Phase noise is incorporated by assuming that the phase of the injected field is a stochastic variable $\phi(\tau)$. In order to insure that the phase will undergo a diffusion process, we consider the variable

$$\delta\omega(\tau) = \dot{\phi} \tag{2}$$

representing the fluctuation of the instantaneous carrier frequency of the input field, $\omega_o + \delta\omega(\tau)$. We assume that $\delta\omega(\tau)$ undergoes a stationary Gaussian process such that

$$\langle\delta\omega(\tau)\rangle = 0, \qquad \langle\delta\omega(\tau)\delta\omega(\tau')\rangle = 2\gamma_c\,\delta(\tau-\tau') \tag{3}$$

where γ_c is the linewidth of the input field (13). As a consequence, the phase undergoes a diffusion process such that (14)

$$\langle\exp[i\phi(\tau)]\rangle = 0,$$

$$\langle\exp[-i\phi(\tau)]\exp[i\phi(\tau)]\rangle = \exp[-\gamma_c|\tau-\tau'|] \tag{4}$$

The inverse of the diffusion constant $\tau_c = \gamma_c^{-1}$ is the coherence time of the driving field; note that also τ_c is measured in units of the inverse atomic linewidth $\gamma_\perp^{-1}$. Our numerical simulation of the phase diffusion process is based on the iterative procedure

$$\phi(\tau_o+\delta\tau) = \phi(\tau_o) + \delta\phi \tag{5}$$

where $\phi(\tau_o)$ is the phase of the incident field at some time τ and $\delta\phi$ is a Gaussian random variable with zero mean, and standard deviation selected in such a way as to fit the required coherence time of the incident light.

3. INSTABILITIES IN THE LASER WITH INJECTED SIGNAL

We focus on parameter values for which the deterministic behavior has already been studied extensively (15,16); they are C=20, Δ=1, θ=2, $\tilde{K}$=0.5, and $\tilde{\gamma}$=0.05. A linear stability analysis shows that, for these parameters, the entire segment from the origin to point A (Fig. 1) of the state equation

$$y = |x|\left[\left(1 - \frac{2C}{1 + \Delta^2 + |x|^2}\right)^2 + \left(\theta + \frac{2C\Delta}{1 + \Delta^2 + |x|^2}\right)^2\right] \tag{6}$$

is unstable against fluctuations. A detailed analysis (15,16) reveals the existence of up to three distinct coexisting attractors for a given value of y in the unstable domain. However, for y<5.58, and y>11.12 there is only one attractor, and in this article we consider only these two domains. For small values of y, the system exhibits regular oscillations with a period of about 2π units of time, corresponding to the beat note between the frequencies $\omega_a=\omega_c$ and ω_o (Fig. 2a). Near the injection locking threshold (point A of Fig. 1), the system displays period-one oscillations with a period of about 4 units of time (Fig. 2c). On decreasing the strength of the driving field, one finds a cascade of period doubling bifurcations that eventually leads to chaotic behavior (Fig. 2e,g). The chaotic nature of the dynamics in this situation has been confirmed with the evaluation of the Lyapunov exponents (16).

4. NUMERICAL RESULTS FOR THE LASER WITH AN INJECTED SIGNAL

The following results are based on the numerical evaluation of very long time series, their power spectra, and Poincare' sections. For y=5, the distortion of the deterministic oscillations is appreciable only when τ_c becomes comparable to the pulsation period, in qualitative agreement with the argument advanced in the introduction (Fig. 3a). The effects induced by the phase diffusion process are especially evident from the Poincare' maps in Fig. (3b,c,d) which have been constructed, as all the other maps shown in this paper, from sequences of 160,000 time steps. Each point in the plane x_1=Re x, x_2=Im x correspond to the crossing of the phase-space trajectory with a preselected plane D=D_o (in this case D=-0.032).

For y=18, the deterministic solution is trongly deformed, even for τ_c as large as 61.7 (Fig. 4a); note that in this case the coherence time is 15 times longer than the period of a single oscillation. Clearly, the effect of phase diffusion is much stronger than in the case y=5, as we can see by direct comparison between the Poincare' maps shown in Fig. 4b,c,d and those of Fig. 3b,c,d. Figures 4e and 4f show two power spectra for different va-

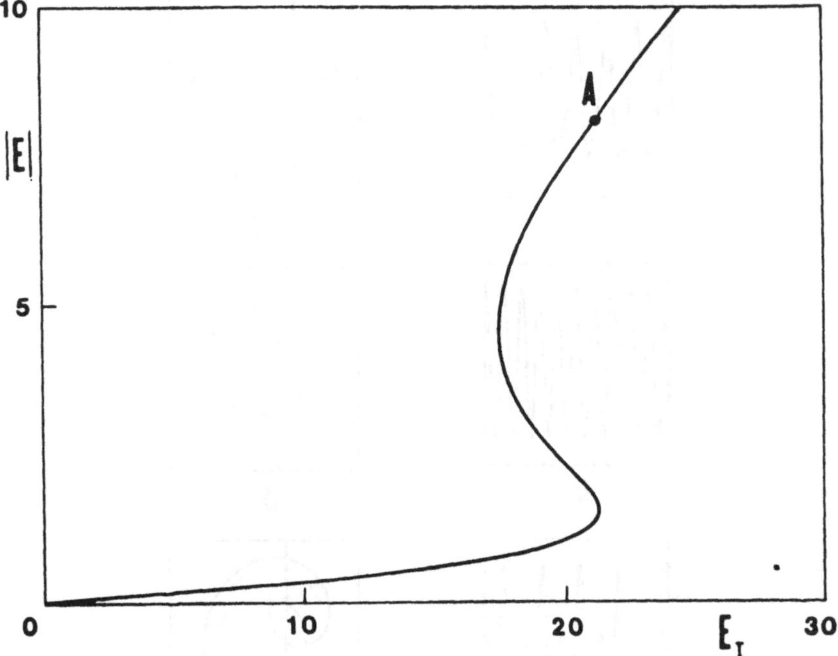

Fig. 1 State equation for a laser with an injected signal corresponding to C=20, Δ=1, θ=2, $\tilde{\kappa}$=0.5, and $\tilde{\gamma}$=0.05. The segment from the origin to point A is unstable against self-pulsing.

lues of τ_c; note that all the spectra displayed in Figs. 4, 6, and 8 are based on the Fourier transform of x_1=Re x and are plotted on a linear scale.

The larger sensitivity to noise for y=18 can be traced, in part, to a geometrical effect. The projection of the phase-space trajectory onto the plane (x_1,x_2) for y=5 is a deformed circle surrounding the origin of the phase-plane (Fig. 2b). Because a transformation of the type $\phi \rightarrow \phi+\delta\phi$ only causes a rotation of the trajectory by a phase angle $\delta\phi$ (or, more precisely, induces a transformation of the type $x \rightarrow x\exp(i\delta\phi)$, $p \rightarrow p\exp(i\delta\phi)$ in the output field and polarization), a circular trajectory around the origin would be entirely unaffected by the phase fluctuation. Thus, for y=5, the phase-induced perturbations are rather small, while for y=18 they are considerably larger; in fact, in the latter case, the phase-space trajectory, whose center is well removed from the origin of the (x_1,x_2) plane, is much more sensitive to phase fluctuations.

For y=14, the period-two deterministic evolution begins to be affected by the phase noise even in a situation where $\tau_c=6.17\times10^3$ (Fig. 5a); for τ_c =6.17 the period-two structure is practically unrecognizable (Fig. 5b). This is confirmed by the corresponding Poincare' maps (Fig. 5c,d) and by the power spectra (Fig. 5e,f).

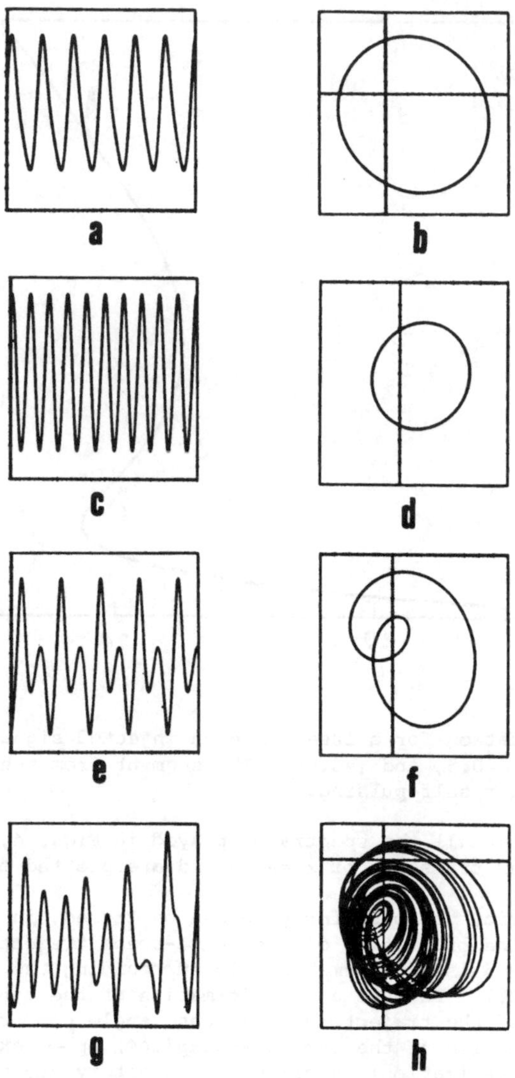

Fig. 2 Curves (a), (c), (e), and (g) show the long-term deterministic evolution of the modulus $|x|$ of the output field for a few values of the input field y. Curves (b), (d), (f), and (h) show the projections of the phase-space trajectory onto the plane x_1=Re x, x_2=Im x; (a) and (b) correspond to y=5; (c) and (d) to y=18; (e) and (f) to y=14; (g) and (h) correspond to chaotic oscillations for y=12.4

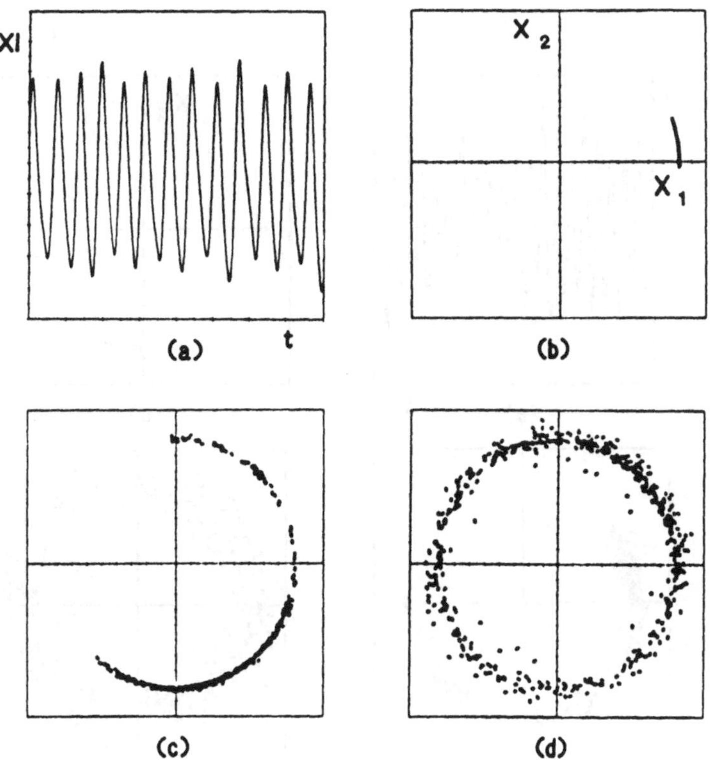

Fig. 3 Effect of phase noise for y=5. (a) Time evolution of $|x|$ for τ_c=5.56; (b), (c) and (d) Poincare' maps for τ_c=6.17x10^3, τ_c=61.7, and τ_c=5.56, respectively.

Figure 6a shows a Poincare' map in the presence of deterministic chaos. This structure, which is already blurred for τ_c=6.17x10^3 (Fig. 6b), disappears completely for τ_c=61.7 (Fig. 6c). On comparing Figs. 6d, e, f, we see that, in the absence of noise the spectrum has a period-two structure on top of a continuous background, while for τ_c=5.56 it is dominated by stochastic noise of the 1/f type.

Of special interest is the problem of the correlation between the phases of the input and output fields (ϕ_{in} and ϕ_{out}, respectively). Plots of the output as a function of the input phase are shown in Fig. 7 using computer runs involving 160,000 temporal iterations. For y=5 the output follows the input phase very closely (Fig. 7a). For y=18 the correlation is weaker (Fig. 7b), but it increases again when the coherence time becomes as small as 5.56.

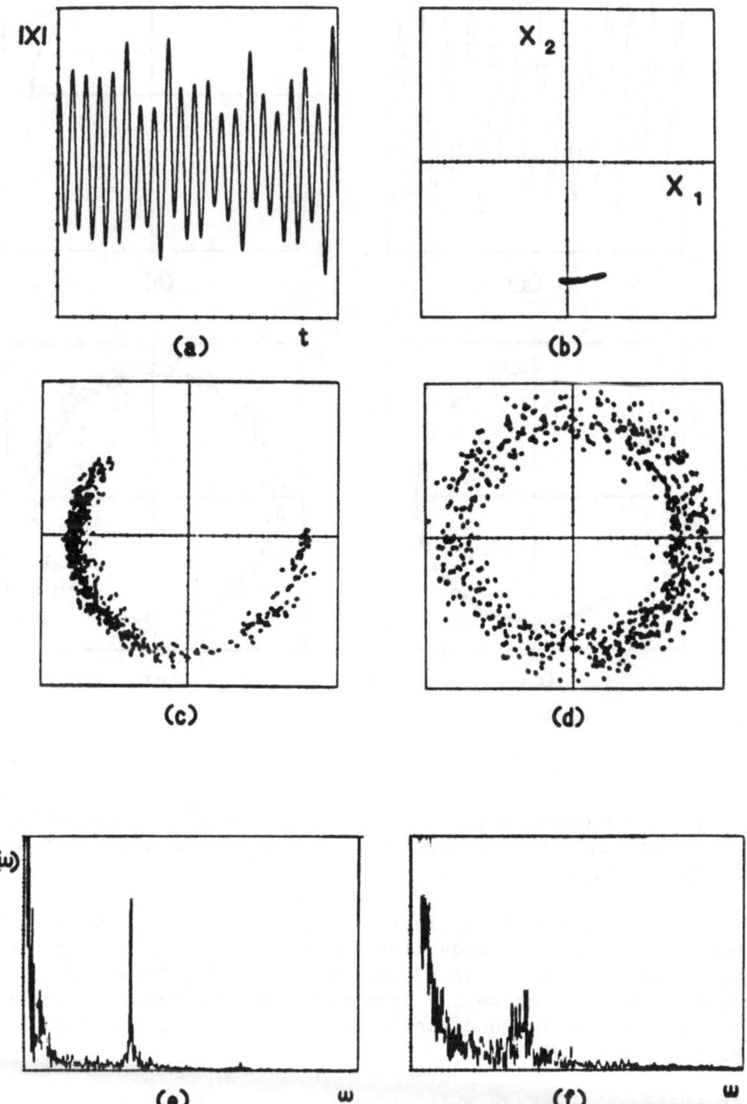

Fig. 4 Effect of phase noise for y=18. (a) Time evolution of $|x|$ for $\tau_c=61.7$; (b), (c) and (d) Poincare' maps for $\tau_c=6.17\times10^3$, $\tau_c=61.7$, and $\tau_c=5.56$, respectively; (e) and (f) power spectrum $S(\omega)$ for $\tau_c=61.7$, and $\tau_c=5.56$, respectively.

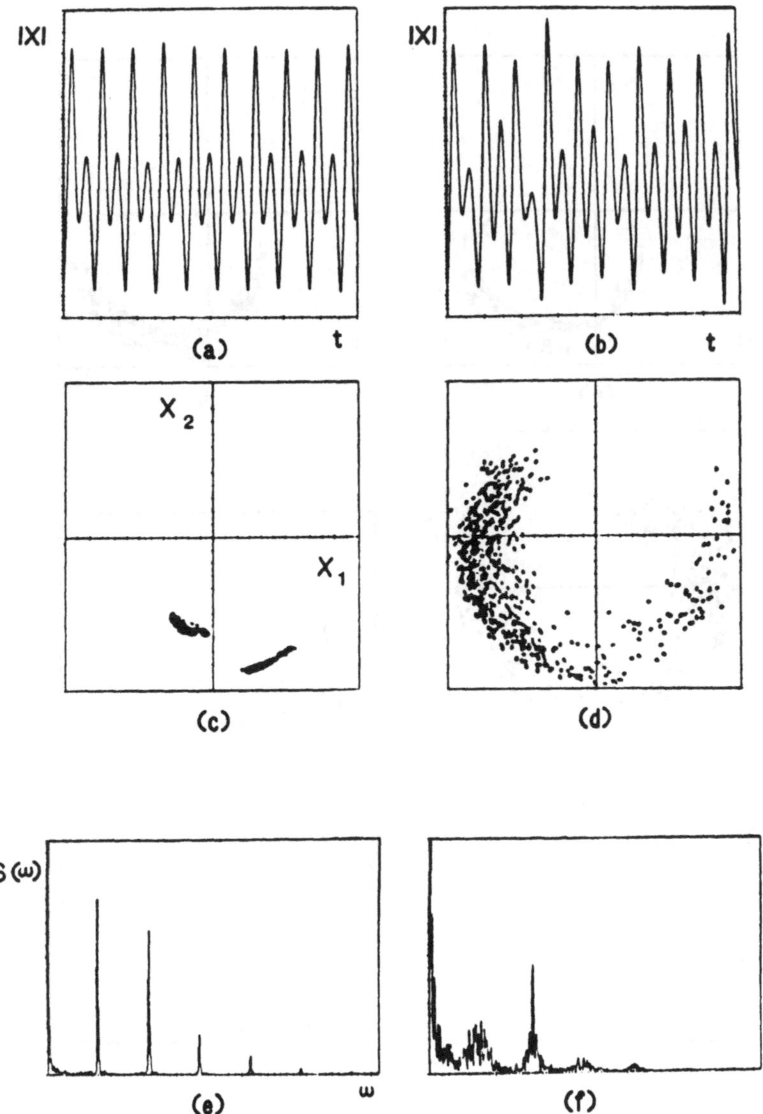

5 Effect of phase noise for y=14. (a) Time evolution of $|x|$ for .17x10^3, and 61.7, respectively; (c) and (d) Poincare' maps for τ_c= x10^3, and 61.7, respectively; (e) and (f) power spectrum $S(\omega)$ for τ_c= x10^3, and 61.7, respectively.

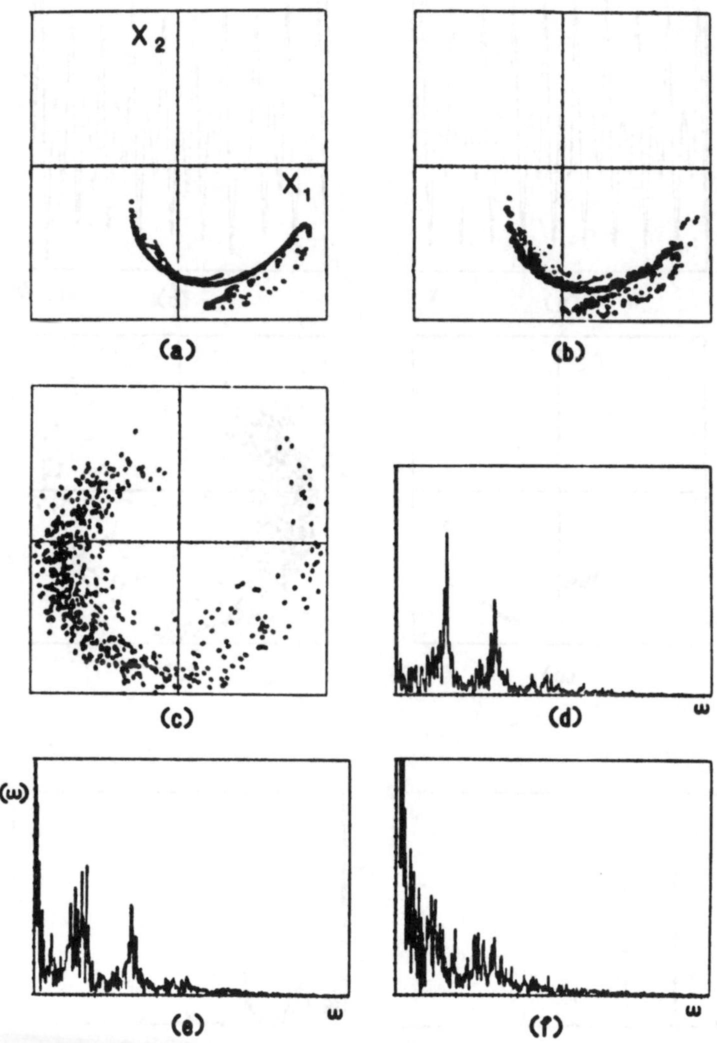

Fig. 6 Effect of phase noise for y=12.4; (a), (b), and (c) Poincare' maps for $\tau_c = \infty$, 6.17×10^3, and 61.7, respectively; (d), (e), and (f) power spectrum for $\tau_c = \infty$, 61.7, and 5.56, respectively.

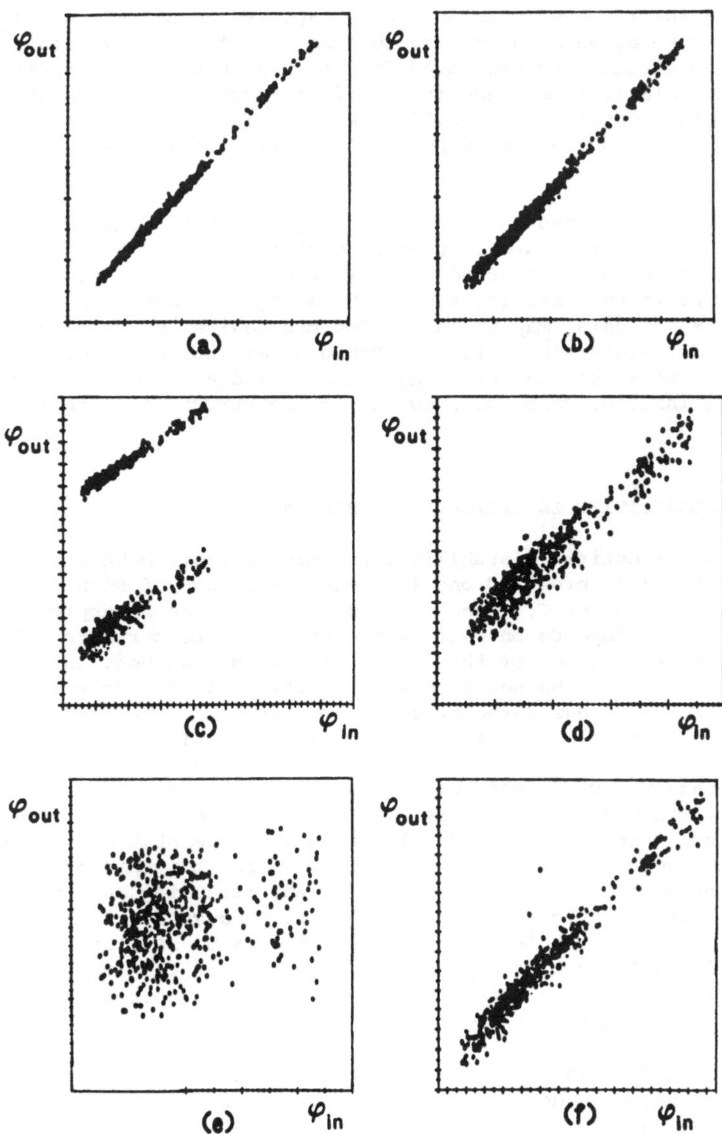

Fig. 7 The phase of the output field is plotted as a function of the phase of the input field for (a) y=5, τ_c=6.17x10^3, (b) y=18, τ_c=6.17x10^3, (c) y=14, τ_c=6.17x10^3, (d) y=14, τ_c=61.7, (e) y=12.4, τ_c=6.17x10^3, (f) y= 12.4, τ_c=5.56.

For y=14 and $\tau_c=6.17\times10^3$ (Fig. 7c), the plot exhibits two distinct regions resulting from the period-two character of the oscillations (Fig. 7c); this structure disappears for $\tau_c=61.7$, as shown in Fig. 7d. However, the correlation between the input and the output phases increases as τ_c decreases, at least over the range of coherence time explored in these runs ($\tau_c>5$). In the cahotic case, when the coherence time is sufficiently long, the output phase is essentially uncorrelated from the input phase, as expected (Fig. 7e); nevertheless, a decrease in τ_c brings about an increase in the degree of correlation until, for $\tau_c=5.56$, the irregular behavior can be ascribed almost entirely to the random phase fluctuations in the input field (Fig. 7f).

We consider finally the case y=30 in the presence of phase noise (Fig. 8a) where the stationary state is stable according to the deterministic model. Figure 8b has been constructed by marking a point in the plane (x_1,x_2) at intervals of 25 temporal iterations. This figure shows a remarkable similarity to the Poincare' map in the period-one regime for y=18 and for the same value of τ_c (see Fig. 4c). The temporal evolution, however, is much more erratic, as we see by comparing Figs. 8a and 4a; the power spectrum (Fig. 8c), instead, does not show any of the structure which is evident in Fig. 4e.

5. INSTABILITIES IN OPTICAL BISTABILITY

In the case of optical bistability, the steady state behavior is still described by Eq. (6), provided one introduces a change of sign in front of the bistability parameter C. When C becomes larger than an appropriate threshold value that depends on $\Delta$and θ, the steady state curve for the output field $|x|$ as a function of the input field y is S-shaped, in a way similar to that of Fig. 1. The negative-slope portion of the curve is always unstable (8). While, under resonant conditions ($\Delta=\theta=0$), the stationary curve is stable along the segment with positive slope, a suitable increase in cavity mistuning creates a competition between the frequency of the input beam ω_o and the nearest cavity resonance; this produces a destabilization of the stationary state along a segment of the upper branch of the steady state curve. This instability can develop even when the stationary curve is not S-shaped. When the bistability parameter is smaller than about 300, the unstable behavior is favored by detuning and mistuning parameters having opposite signs. Typically,Δ should not exceed a value of about one, but θ can be large (for example, $\theta=25$). The spontaneous oscillations that arise from the instability have a period of the order of the cavity build-up time κ^{-1}; oscillations of this type have been detected experimentally with parallel beams of atomic sodium (9).

The model based on Eqs. (1) predicts that for C>300 the output intensity can display higher order temporal structures such as, for example, period-two, period-four and chaos.

6. NUMERICAL RESULTS FOR OPTICAL BISTABILITY

We consider first the parameters C=100, $\Delta=0$, $\theta=-9$, $\tilde{\kappa}=0.3$, and $\tilde{\gamma}=1.6$; in this case the unstable domain corresponds to the range $61.9<y<87.5$. We consider two values of τ_c; the first, $\tau_c=538$, is meant to simulate the linewidth of the driving field in the experiments of Ref. 9; the second, $\tau_c=61.7$, is used to compare the response of optical bistability with that of

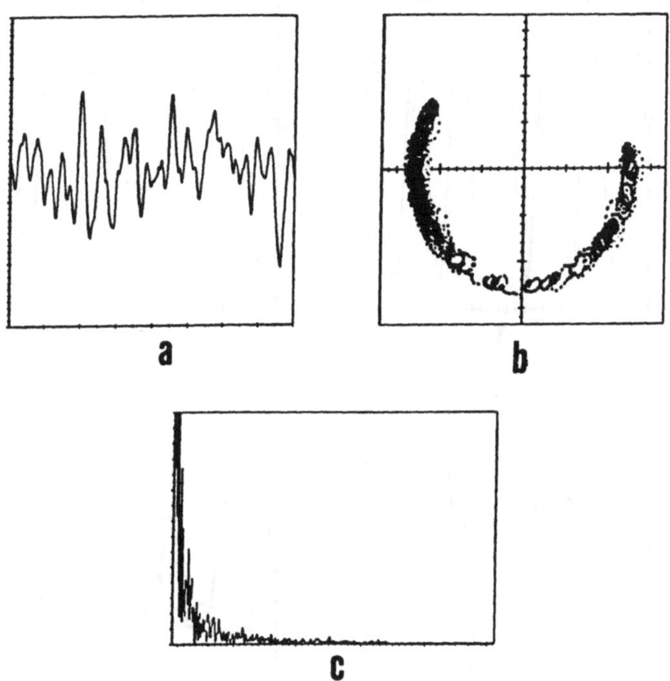

Fig. 8 Effect of phase noise for y=30, τ_c=61.7; (a) long-term time evolution of the modulus $|x|$ of the output field, (b) phase-space diagram, (c) power spectrum.

the laser with an injected signal.

For y=80, the period of deterministic oscillations is about 2 units of time. The influence of phase noise is quite small for τ_c=538, as we see in Fig. 9a; for τ_c=61.7, it is still considerably smaller than in the corresponding case of the LIS (compare Figs. 9b and 4a). Here, qualitatively, the noise sensitivity is about the same as observed in the small-y range of the LIS (e.g. y=5), in spite of the fact that the projection of the phase-space trajectory on the (x_1, x_2) plane is centered away from the origin.

For y=105, the deterministic theory predicts a stable stationary state. As shown in Figs. 9c and d, and just as noted with the laser with an injected signal, the injection of noise produces irregular oscillatory behavior in the output signal. The presence of these oscillations may be an obstacle in attempting a quantitative comparison between the experimental instability threshold and its theoretical counterpart.

Next we consider the parameters C=400, Δ= -1, θ=20, $\tilde{\kappa}$=0.4, $\tilde{\gamma}$=1.76, and y= 297.3; in this case the output exhibits chaotic behavior with a fundamental period slightly smaller than one, as shown in Fig. 10. The spectrum in

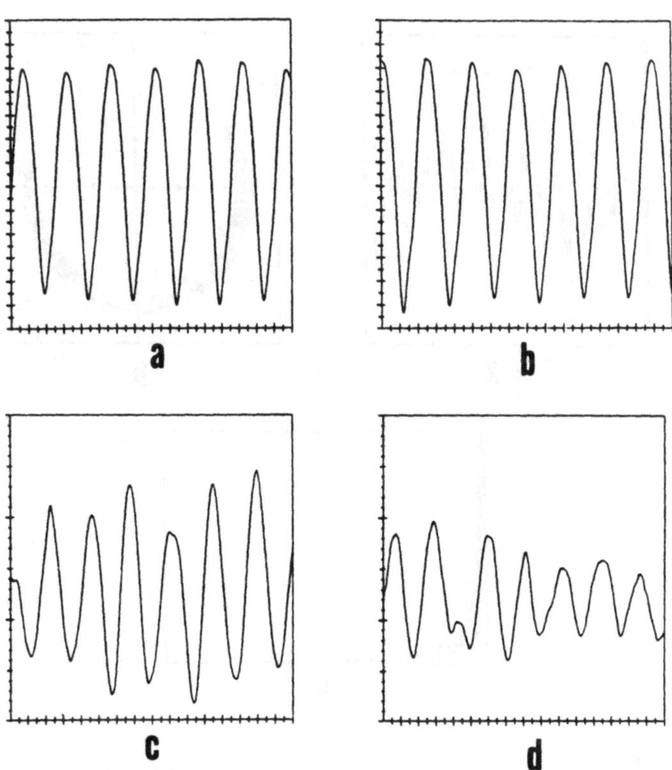

Fig. 9 Optical bistability. Snapshots of the long-time evolution for C=100, Δ=0, θ=-9, $\tilde{\kappa}$=0.3, $\tilde{\gamma}$=1.6, and (a) y=80, τ_c=538, (b) y=80, τ_c=61.7, and (c) and (d) y=105, τ_c=538.

Fig. 10d (the Fourier transform of x_1=Re x) displays a period-four structure riding on a continuous background. Here the discrete spectral components are much higher than in the case of the LIS considered before, and for this reason this spectrum is plotted on a semilogarithmic scale.

The effect of phase noise is illustrated in Fig. 11. Again, we find that OB is less sensitive to this type of disturbance than the LIS. A direct comparison of Fig. 11 and Fig. 6c shows that the Poincare' map is less blurred in the case of optical bistability, and some signatures of the original deterministic map ($\tau_c=\infty$) shown in Fig. 10c still persist for τ_c=61.7. The plot of the output as a function of the input phase, displayed in Fig. 11e, shows again a marked period-four character in the chaotic oscillations. This feature persists to some extent even for a short coherence time (τ_c= 5.56) as shown in Fig. 11f.

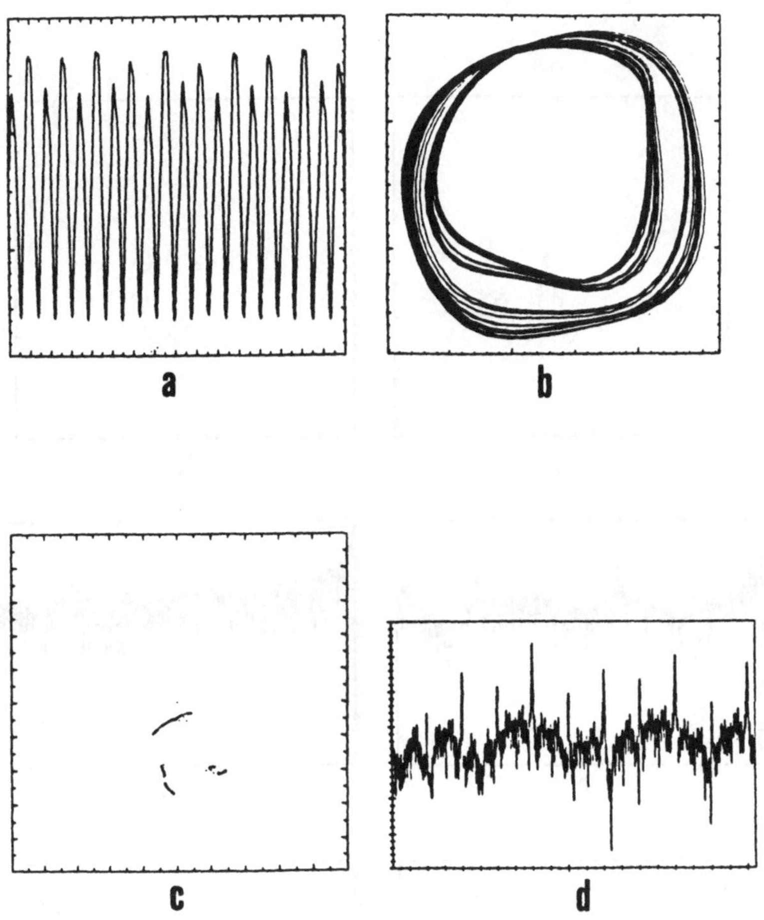

Fig. 10 Optical bistability for $C=400$, $\Delta=-1$, $\theta=20$, $\tilde{\kappa}=0.4$, $\tilde{\gamma}=1.76$, $y=297.3$, and $\tau_c=\infty$ (no phase noise); (a) time evolution of $|x|$; (b) projection of the phase-space trajectory onto the plane $x_1=\mathrm{Re}\ x$, $x_2=\mathrm{Im}\ x$; (c) Poincare' map; (d) power spectrum on a semi-logarithmic scale.

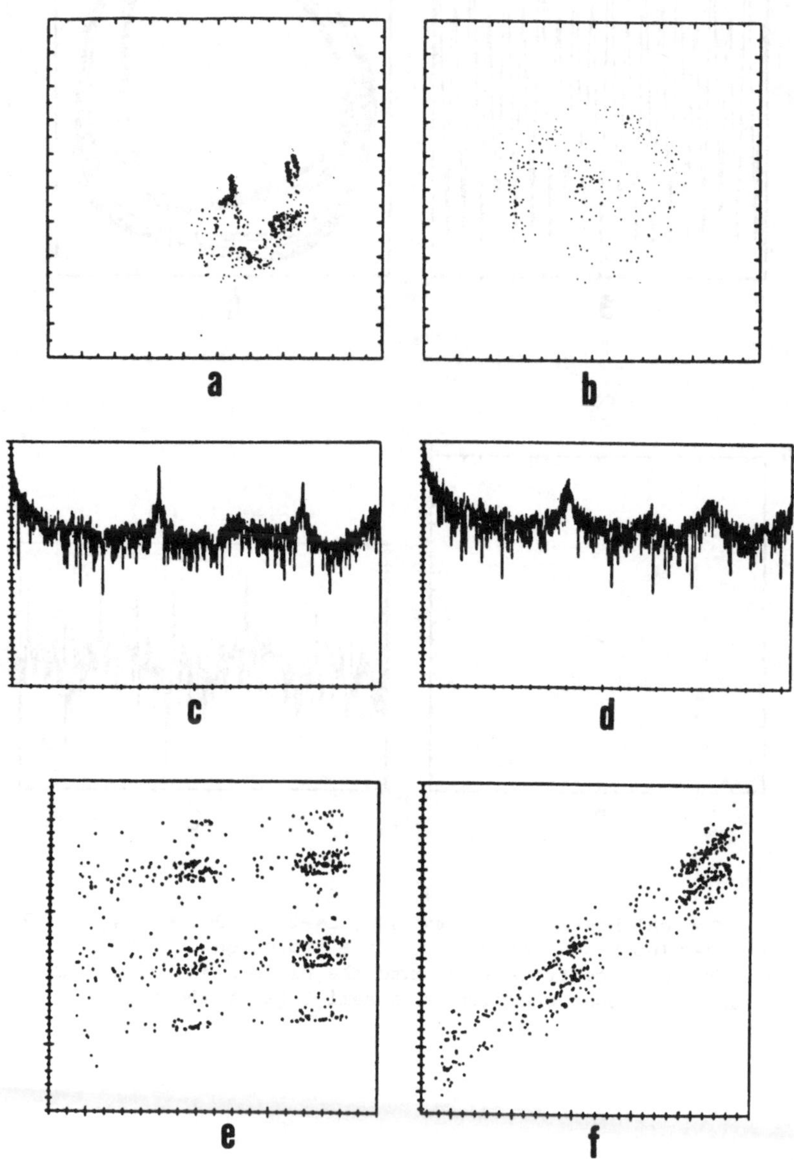

Fig. 11 Optical bistability. Except for τ_c the parameters are the same as in Fig. 10; (a),(b) Poincare' maps for τ_c=61.7 and τ_c=5.56; (c) and (d) power spectrum for τ_c=61.7 and τ_c=5.56; (e) and (f) output field phase as a function of the input field phase for τ_c=6.17x10^3 and τ_c=5.56.

REFERENCES

1. J.P. Crutchfield, J.D. Farmer, and B.A. Huberman, Phys. Rept. 92, 45 (1982).
2. M. Brambilla, L.A. Lugiato, G. Strini, and L.M. Narducci, in Coherence Cooperation and Fluctuations, edited by F. Haake, L.M. Narducci, and D.F. Walls (Cambridge University Press, Cambridge, 1986) p.185.
3. M. Brambilla, L.A. Lugiato, G. Strini, and L.M. Narducci, Phys. Rev. A34, 1237 (1986).
4. R.J. Glauber, in Quantum Optics and Electronics, Les Houches Summer School (Gordon and Breach, New York, 1964), p. 63.
5. H. Haken, Handbuch der Physik, Vol XXXV/2c (Springer Verlag, Berlin, 1970).
6. M. Sargent, III, M.O. Scully, and W.E. Lamb, Jr., Laser Physics (Addison-Wesley, New York, 1974)
7. M.B. Spencer and W.E. Lamb, Jr., Phys. Rev. A5, 884 (1972). See, also, J. Opt. Soc. Am. B2, January, 1985; special issue on Instabilities in Active Optical Media, edited by N.B. Abraham, L.A. Lugiato, and L.M. Narducci.
8. See, for example, L.A. Lugiato, in Progress in Optics, Vol. XXI, edited by E. Wolf (North Holland, Amsterdam, 1984), p. 69.
9. L.A. Orozco, A.T. Rosemberger, H.J. Kimble, Phys. Rev. Lett. 53, 2547 (1984).
10. E. Brun, B. Derighetti, M. Ravani, G. Broggi, P. Meier, R. Stoop, and R. Badii, in Proceedings of the 1986 New Zealand School in Quantum Optics, edited by J.D. Harvey and D.F. Walls (Springer Verlag, Berlin, 1986), in press.
11. J.L. Boulnois, A. VanLerberghe, P. Cottin, F.T. Arecchi, and G.P. Puccioni, Opt. Comm. 58, 124 (1986).
12. L.A. Lugiato, L.M. Narducci, D.K. Bandy, and C.A. Pennise, Opt. Comm. 46, 64 (1983).
13. G.S. Agarwal, Phys. Rev. Lett. 37, 1383 (1976).
14. J.H. Eberly, Phys. Rev. Lett. 37, 1387 (1976).
15. (a) D.K. Bandy, L.M. Narducci, C.A. Pennise, and L.A. Lugiato, in Coherence and Quantum Optics V, edited by L. Mandel and E. Wolf (Plenum Press, New York, 1984)p. 585; (b) D.K. Bandy, L.M. Narducci, and L.A. Lugiato, J. Opt. Soc. Am. B2, 148 (1985), and references quoted therein.
16. Y. Gu, D.K. Bandy, J.M. Yuan, and L.M. Narducci, Phys. Rev. A31, 354 (1985).

CHAOS IN THE MICROMASER

P MEYSTRE AND E M WRIGHT

1. INTRODUCTION

"Cavity quantum electrodynamics" is an area of research that has gained considerable importance in the last few years. It can be defined as the study of the interaction between one or a few atoms or electrons and an electromagnetic field with mode structure and density of states significantly different from those of free space. Such fields can be found in high-Q microwave cavities or in electron traps. Investigations along these lines have allowed the observation of order-of-magnitude effects attributable to the quantized nature of the electromagnetic field, such as the modification of the spontaneous lifetime of a transition in a tailored electromagnetic field environment (Goy et al., 1983; Hulet and Kleppner, 1985; Gabrielse and Demhelt, 1985).

An extreme situation is reached when a single atom interacts with the field of a single-mode lossless resonator. The theoretical description of this system is given by the well-known Jaynes-Cummings model (Jaynes and Cummings, 1963). Over the years a number of predictions of this model were analyzed, including the so-called "Cummings collapse" (Cummings, 1965), as well as "revivals" (Eberly et al., 1980) whose existence is a signature of the truly quantum-mechanical nature of the electromagnetic field (for a recent review see Barnett et al. (1986)). Experiments on Rydberg atoms in extremely high-Q cavities now approach the conditions to check those predictions (Meschede et al., 1985) and indeed Cummings collapse and revivals have recently been observed in the laboratory (Walther, 1985). What was once thought to be a rather unrealistic theorists' testing ground for basic ideas on the foundations of quantum optics is now within the reach of experiments.

In the present paper we discuss novel aspects of cavity QED that are not only relevant to traditional quantum optics (lasers and masers, coherence theory, non-classical fields, etc.), but simultaneously make contact with nonlinear dynamics and chaos (see e.g., H. G. Schuster, 1984). We consider specifically a micromaser (Meschede et al., 1985; Filipowicz et al., 1986a) consisting of a single-mode cavity in which two-level atoms are injected at such a low rate that at most one atom at a time is present inside the cavity. This is of course a problem that has been widely studied in quantum optics in the context of maser theory. However, in the present case we do not perform the usual approximations associated with the operation of a maser. In particular, since we consider a single-mode cavity, the irreversible spontaneous emission which is usually effective in smoothing out the quantum mechanical phases in the atom-field interaction is ineffective (Filipowicz et al., 1986b).

This paper concentrates on the semi-classical version of this model and shows that it exhibits chaos under a rather general set of circumstances. The semi-classical micromaser presents a number of coexisting attractors. In the generic case, each attractor follows a period-doubling route to chaos, with a subsequent inverse route back to stability, as one order parameter of the system (e.g., the cavity losses) is varied. But the various bifurcations occur at different values of the order parameter for different attractors. Also, because of the complexity of the system return map, other possibilities such as intermittency and crises can not be ruled out in general. In view of these results, it is expected that a future study of the fully quantized version of the micromaser will help shed light on the problem of quantum chaos.

The rest of this paper is organized as follow: Section 2 reviews results of a lossless, semi-classical micromaser. Filipowicz et al. (1986c) showed that this model system does not exhibit chaos, although it displays a sensitive dependence on initial conditions. Specifically, as successive atoms are injected the system evolves towards one of infinitely many fixed points. These fixed points correspond to values of the field such that each atom undergoes a $2q\pi$ coherent interaction (q integer) during its passage through the resonator.[8] They are only marginally stable, i.e., stable when approached from one side but unstable when approached from the other, depending on the initial atomic inversion. Section 3 generalizes these results to include losses. As in Filipowicz et al. (1986b), we simplify the description by neglecting the cavity losses while an atom resides inside the cavity. This approximation is discussed in detail by Barnett and Knight (1985) who show that it is very good under the conditions of the Garching experiment (Meschede et al., 1985). Section 3 shows that in contrast to the lossless case, adding cavity damping almost invariably leads to a chaotic response of the system. But in general the micromaser has several coexisting fixed points whose basins of attraction are rather intricately entwined, with a sensitive dependence on the initial conditions. Finally, Section 4 is a conclusion and outlook on future developments.

2. REVIEW OF THE LOSSLESS MODEL

We consider first the situation where a monoenergetic beam of atoms is injected into a lossless cavity, such that at most one atom at a time interacts with the field (Filipowicz et al., 1986c). The interaction of a single atom with the field is given in the rotating wave approximation by the Jaynes-Cummings Hamiltonian (Jaynes and Cummings, 1963)

$$\mathcal{H} = (\hbar\omega_0/2)S_3 + \hbar\omega a^{\dagger}a + \hbar\kappa/2\,(S_+a + a^{\dagger}S_-)\ , \tag{1}$$

where ω_0 is the energy difference between the two atomic levels, κ the electric dipole coupling constant, a and $a^{\dagger}$ the field annihilation and creation operators and ω its frequency, and S_3, $S_{\pm}$ the Pauli spin operators. The exact time evolution of this system is well-known. The classical version of the model is obtained by factorizing the expectation values in the Heisenberg equations of motion for the atomic and field operators (see e.g. Allen and Eberly, 1975). This gives

$$\dot{\sigma}_- = i\Delta\sigma_- + i(\kappa/2)\,\sigma_3 E\ , \tag{2a}$$

$$\dot{\sigma}_3 = -i\kappa(\sigma_+E - E^*\sigma_-)\ , \tag{2b}$$

$$\dot{E} = -i(\kappa/2)\,\sigma_-\ , \tag{2c}$$

and $\sigma_+ = \sigma_-^*$, where $\Delta = \omega - \omega_0$ is the atom-field detuning and

$$E = \langle a\rangle\exp(i\omega t)\ , \tag{3a}$$

$$\sigma_- = \langle S_-\rangle\exp(i\omega t)\ , \tag{3b}$$

$$\sigma_3 = \langle S_3\rangle\ . \tag{3c}$$

The quantum version of this model is discussed in Filipowicz et al. (1985c). Here, we concentrate on the semi-classical version, which will then be generalized in Section 3 to the case of a lossy cavity. We focus on the resonant case, and for convenience choose the phase of the classical field that only couples to the out-of-phase component of the polarization. In terms of the length R of the Bloch vector and the tipping angle θ defined through the equations

$$i(\sigma_+ - \sigma_-) = R\sin\theta\ , \tag{4a}$$

$$\sigma_3 = R\cos\theta \ , \tag{4b}$$

and with the dimensionless time $\xi = \kappa t$, Eqs. (2) become

$$\frac{\partial R}{\partial \xi} = 0 \ , \tag{5a}$$

$$\frac{\partial \theta}{\partial \xi} = E \ , \tag{5b}$$

$$\frac{\partial E}{\partial \xi} = (R/4)\sin\theta \ . \tag{5c}$$

The atoms are assumed to be injected inside the resonator without initial dipole moment, $\theta = 0$. We take advantage of the conservation of the magnitude R of the Bloch vector, $R = \sigma_3(t=0)$ to carry out the further transformation

$$\tau = \xi\sqrt{R}/2 \ , \tag{6a}$$

$$\mathcal{E} = \frac{2E}{\sqrt{R}} \ . \tag{6b}$$

The semiclassical model then reduces to the pendulum equations

$$\frac{\partial \theta}{\partial \tau} = \mathcal{E} \ , \tag{7a}$$

$$\frac{\partial \mathcal{E}}{\partial \tau} = \eta\sin\theta \ , \tag{7b}$$

where $\eta \equiv \mathrm{sign}(R)$. The sign of the inversion alone determines the qualitative behavior of the model.

Let us denote by $\mathcal{E}_n$ the dimensionless intracavity field before injection of atom n+1. Depending upon whether the atom is initially inverted or not, the Bloch vector starts in upright or downright position with angular velocity $\mathcal{E}_n$. The field $\mathcal{E}_{n+1}$ at the instant of escape of this atom is obtained by integrating Eqs. (7) over the dimensionless interaction time τ_{int}. This yields the return map

$$\mathcal{E}_{n+1} = \mathcal{F}(\mathcal{E}_n). \tag{8}$$

The function $\mathcal{F}(\mathcal{E})$, which also depends on the control parameters η and τ_{int}, can be implicitly obtained by taking advantage of the energy conservation relation

$$\frac{\dot{\theta}^2}{2} + \eta\cos\theta = \frac{\mathcal{E}_n{}^2}{2} + \eta \ , \tag{9}$$

which, when inserted into Eq. (7b), yields

$$\int_{\mathcal{E}_n}^{\mathcal{F}(\mathcal{E}_n)} d\mathcal{E} \left[1 - \left\{ \frac{\mathcal{E}_n{}^2 - \mathcal{E}^2}{2} + \eta \right\}^2 \right]^{(-1/2)} = \eta\tau_{int} \quad . \tag{10}$$

If the intracavity field $\mathcal{E}_n$ before injection of atom n+1 is large enough and the field consequently varies only little under the influence of one atom, Eq. (9) gives approximately

$$\mathcal{E}_{n+1} \cong \mathcal{E}_n + (2\eta/\mathcal{E}_n)\sin^2(\mathcal{E}_n\tau_{int}/2) \ . \tag{11}$$

But in general, the easiest way to obtain the map $\mathcal{F}$ is to integrate numerically the pendulum Eqs. (7) directly.

Figure 1 shows the return map $\mathcal{E}_{n+1} = \mathcal{F}(\mathcal{E}_n)$ for the case of initially inverted atoms ($\eta > 0$) and $\tau_{int} = 5$. The evolution of the field as successive atoms are injected inside the resonator is obtained by successive iterates of the map. Since for $\eta > 0$ the function $\mathcal{F}$ satisfies $\mathcal{F}(\mathcal{E}) \geq \mathcal{E}$ and its oscillations decrease in amplitude for increasing $\mathcal{E}$ (see Eq. (11)); the iteration always converges towards one of the fixed points $0 = \mathcal{E}^*_0 < \mathcal{E}^*_1 < .. < \mathcal{E}^*_j < ...$ given by the solutions of $\mathcal{E}^* = \mathcal{F}(\mathcal{E}^*)$ (Filipowicz et al., 1986c).

The fixed point $\mathcal{E}^*_q$ clearly corresponds to a $2q\pi$ field-atom interaction: for this value of the intracavity field, the interaction time τ_{int} is precisely such that the Bloch vector performs a $2q\pi$ rotation, leaving the total system unchanged when the atoms leave the cavity. The coherent nonlinear nature of the semiclassical atom-field interaction dynamically prevents the unlimited growth of the field even in the absence of damping.

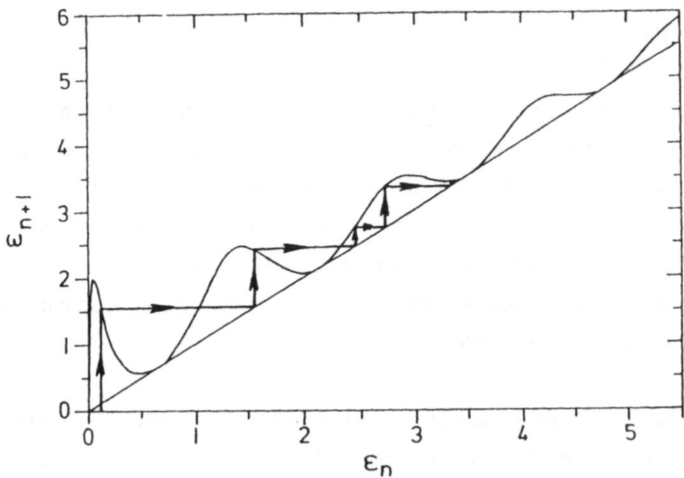

Figure 1. Return map $\mathcal{E}_{n+1} = \mathcal{F}(\mathcal{E}_n)$ for initially inverted atoms and $\tau_{int} = 5$.

Because the map $\mathcal{F}(\mathcal{E})$ is not reversible, this leads to the construction of tightly interwoven basins of attraction of the various fixed points, with a concomitant sensitive dependence on initial conditions. The iterates of the mapping $\mathcal{E}_{n+1} = \mathcal{F}(\mathcal{E}_n)$ are "almost chaotic": Although any initial point always leads asymptotically to one of the fixed points $\mathcal{E}^*_j$, which one it will be sensitively depends on the initial value $\mathcal{E}_0$ of the field. For completeness, we note that the inherent quantum fluctuations present in a quantized field almost always force it eventually to grow past the semi-classical fixed points. A notable exception occurs however if the field evolves towards a number state with no intensity fluctuations (Filipowicz et al., 1986c).

3. LOSSY RESONATOR

In this section we generalize the preceeding results to the case of a lossy resonator. For simplicity, we however assume that cavity losses can be ignored while an atom is inside the cavity (Filipowicz et al., 1986b). Thus, the dynamics of the system are now described by the map

$$\mathcal{E}_{n+1} = \alpha\mathcal{F}(\mathcal{E}_n) \equiv \mathcal{D}(\mathcal{E}_n) \ , \tag{12}$$

where the attenuation coefficient

$$\alpha = \exp(-\gamma\tau_0) \tag{13}$$

depends only on the cavity decay rate γ and the interval τ_0 between atoms, taken here to be constant. (Note that this assumption was not needed in the preceeding section, but is crucial when cavity losses are introduced. Introducing fluctuations in τ_0 can significantly change the behavior of the system, as discussed by Filipowicz et al., 1986b). $\mathscr{E}_{n+1}$ is now the intracavity field just before injection of atom n+2. In Section 2, it was the same as the field at the instant of escape of atom n+1, but because of damping, we must now distinguish between these two fields.

The return map $\mathscr{D}$ is illustrated in Fig. 2 for an attenuation coefficient $\alpha = 0.9$ and $\tau_{int} = 9$. It is of the same form as $\mathscr{F}$, except that the local minima lie now on the straight line $\mathscr{E}_{n+1} = \alpha\mathscr{E}_n$. This seemingly small difference has drastic consequences.

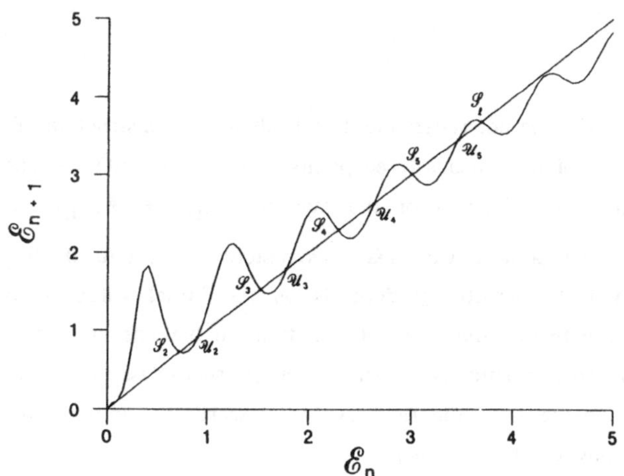

Figure 2. Return map $\mathscr{E}_{n+1} = \mathscr{D}(\mathscr{E}_n)$ for initially inverted atoms, $\tau_{int} = 9$, and $\alpha = 0.9$. $\mathscr{S}_i$ and $\mathscr{U}_i$ label the fixed points of the map, see text for details.

Instead of an infinite set of marginally stable fixed points $\mathscr{E}^*_j$, the map now possesses a *finite* number of fixed points with an alternance of absolutely unstable fixed points $\mathscr{U}_j$ and of *potentially* unstable fixed points $\mathscr{S}_j$ which can evolve into more complex attractors $\mathscr{A}_j$. The last fixed point $\mathscr{S}_\ell$ is always potentially unstable. As in the lossless case, the map is not invertible, and determining the basins of attraction of the various fixed points requires a detailed numerical analysis. In fact, our preliminary study cannot rule out that the boundary between different basins of attraction is itself a fractal, as was found in the driven damped pendulum (Gwinn and Westervelt, 1986). However, a few comments of a general nature can already be made by simple inspection of the map.

(i) Let us label $\mathcal{E}^i_{max}$ the local maximum of the return map between the fixed points $\mathcal{U}_{i-1}$ and $\mathcal{S}_i$. There is always a value $i = k$ such that $\mathcal{E}^k_{max} < \mathcal{U}_k$. No initial condition below $\mathcal{E}^k_{max}$ can evolve past this point.

(ii) All initial fields larger than $\mathcal{S}_\ell$ belong to the basin of attraction of this fixed point, or more generally of the attractor $\mathcal{A}_\ell$.

(iii) Because the slope of the return map $\mathcal{D}(\mathcal{S}_i)$ varies with i, so does the stability of these fixed points. This implies that situations where, say, $\mathcal{S}_1$ is stable but $\mathcal{S}_j$ becomes a chaotic attractor, $\mathcal{A}_j$ can be expected.

Such a situation occurs in particular in the example of Fig. 2. In this case, the fixed points $\mathcal{S}_1$ and $\mathcal{S}_2$ are stable, but $\mathcal{S}_3$, $\mathcal{S}_4$, and $\mathcal{S}_5$ are unstable and replaced by a period-4 attractor $\mathcal{A}_3$, a quasi-periodic attractor $\mathcal{A}_4$, and a period-2 attractor $\mathcal{A}_5$, respectively. Finally, the last fixed point $\mathcal{S}_6 \equiv \mathcal{S}_\ell$ is stable. These results are summarized in Fig. 3, which shows iterations of the return map $\mathcal{D}$ in the vicinity of $\mathcal{S}_2$ to $\mathcal{S}_5$. Figure 4 shows the inversion of the two-level atoms exiting the resonator as a function of the number of iterations in the case of the period-4 attractor $\mathcal{A}_3$.

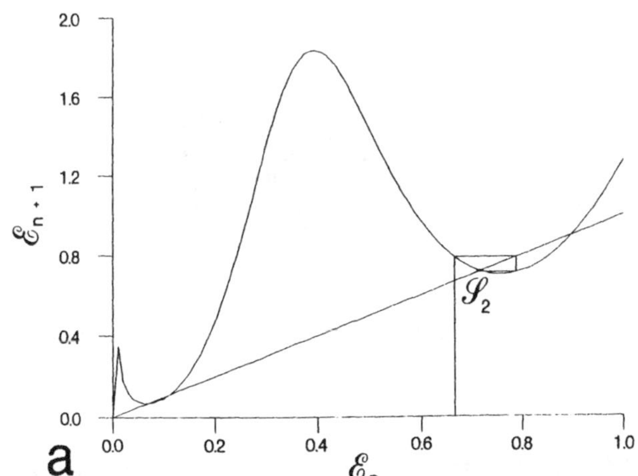

Figure 3. Iterates of the return map $\mathcal{D}$ for $\alpha = 0.9$ and $\tau_{int} = 9$ illustrating (a) the approach to the fixed point $\mathcal{S}_2$; (b) the period-4 attractor $\mathcal{A}_3$ about $\mathcal{S}_3$; (c) the quasi-periodic attractor $\mathcal{A}_4$; (d) the period-2 attractor $\mathcal{A}_5$. *(Figure continued next page.)*

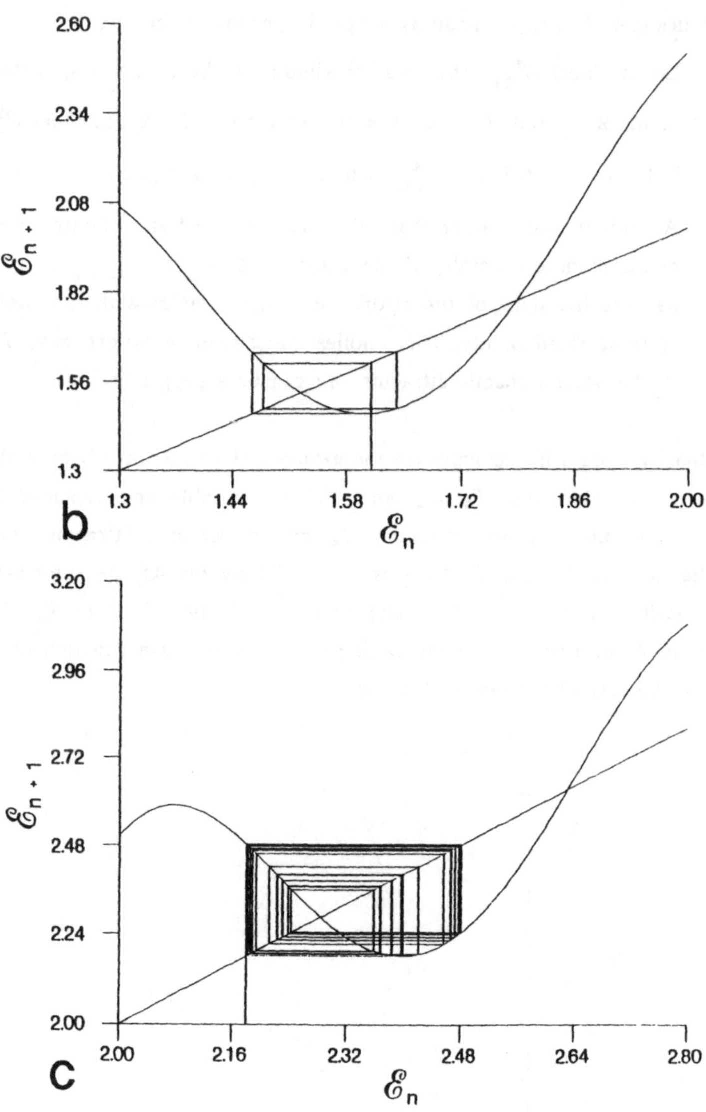

Figure 3 (*continued*). Iterates of the return map $\mathscr{D}$ for $\alpha = 0.9$ and $\tau_{\text{int}} = 9$: (b) the period-4 attractor $\mathscr{A}_3$ about $\mathscr{S}_3$; (c) the quasi-periodic attractor $\mathscr{A}_4$. (*Figure continued next page.*)

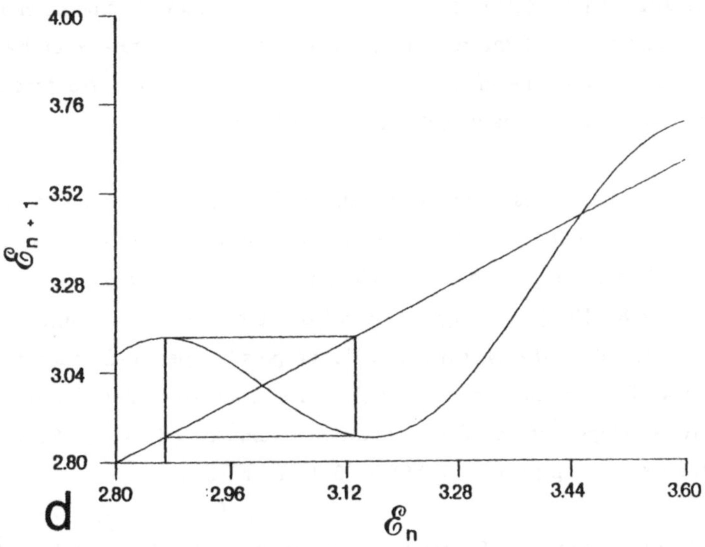

Figure 3 (*continued*). Iterates of the return map $\mathscr{D}$ for $\alpha = 0.9$ and $\tau_{int} = 9$: (d) the period-2 attractor $\mathscr{A}_5$.

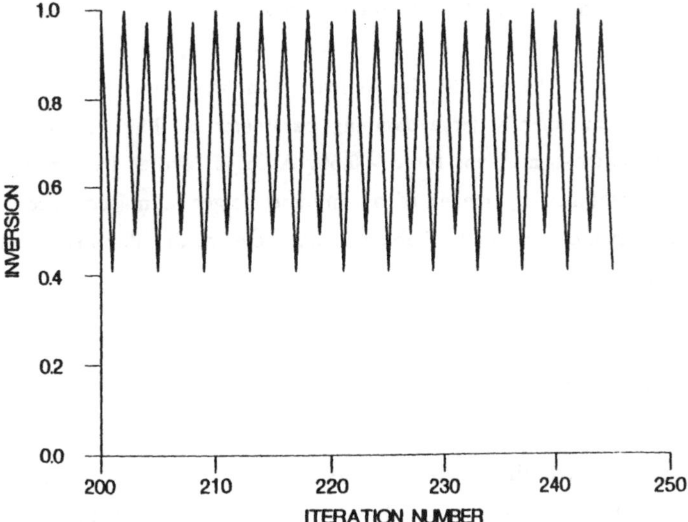

Figure 4. Period-4 behavior of the atomic inversion σ_3 at the exit of the cavity in the attractor $\mathscr{A}_3$. Same parameters as in Fig. 2.

While remaining in the vicinity of one of the fixed points $\mathcal{S}_i$, one (or more) parameter of the micromaser can be varied, leading to a series of bifurcations to chaos. In this paper, we vary the attenuation coefficient α. This could be achieved experimentally either by changing the temperature of the resonator, and hence its decay rate γ, or by changing the atomic beam density and the interval τ_0 between atomic arrivals. We concentrate on the region of the return map in the vicinity of the fixed point $\mathcal{S}_3$.

For a lossless system, $\mathcal{S}_3$ is a marginally stable fixed point. Its stability first increases when increasing α, but for $\alpha \cong 0.93$, we find that a bifurcation to a period-2 attractor has taken place. The system goes then through what is most likely a Feigenbaum sequence (Feigenbaum, 1978, 1979), although a detailed numerical check still remains to be performed. For $\alpha = 0.88$, the system is chaotic (or possibly period-2 chaos, as indicated by the Fourier transform of the atomic inversion after successive iterates) and for $\alpha = 0.8$ it exhibits fully developed chaos. These results are summarized in Fig. 5a-d, which shows iterates of the return map $\mathcal{D}$ for $\alpha = 0.95$, 0.93, 0.88 and 0.80.

Figure 6a shows a sequence of iterates of the atomic inversion and Fig. 6b its Fourier transform for the case $\alpha = 0.80$. Although the Liapounov exponent will have to be computed to determine with absolute certainty that the system is chaotic, both the broadband Fourier transform and the local nature of the return map leave little doubt that the system is following a Feigenbaum route to chaos.

A simple inspection of the map $\mathcal{D}$ also indicates that the Feigenbaum sequence has to be interrupted for a sufficiently large damping parameter α. This is illustrated in Fig. 5e, which shows a period-8 attractor for $\alpha = 0.60$. As α is further increased, the fixed point $\mathcal{S}_3$ recovers its stability by means of an inverse Feigenbaum sequence, and eventually disappears through a crisis. Similar behaviors characterize the stability of all unstable-type fixed points $\mathcal{S}_i$.

4. CONCLUSIONS

In this paper, we have illustrated the way in which a semi-classical version of the micromaser is generally chaotic and exhibits a number of coexisting basins of attraction. Our results are still preliminary, and a detailed numerical study is under way which will give a complete and detailed description of this system. In particular, our model exhibits a strong analogy with the driven damped pendulum (Gwinn and Westervelt, 1986), except that the driving of the pendulum is produced in our case by a sequence of coherent,

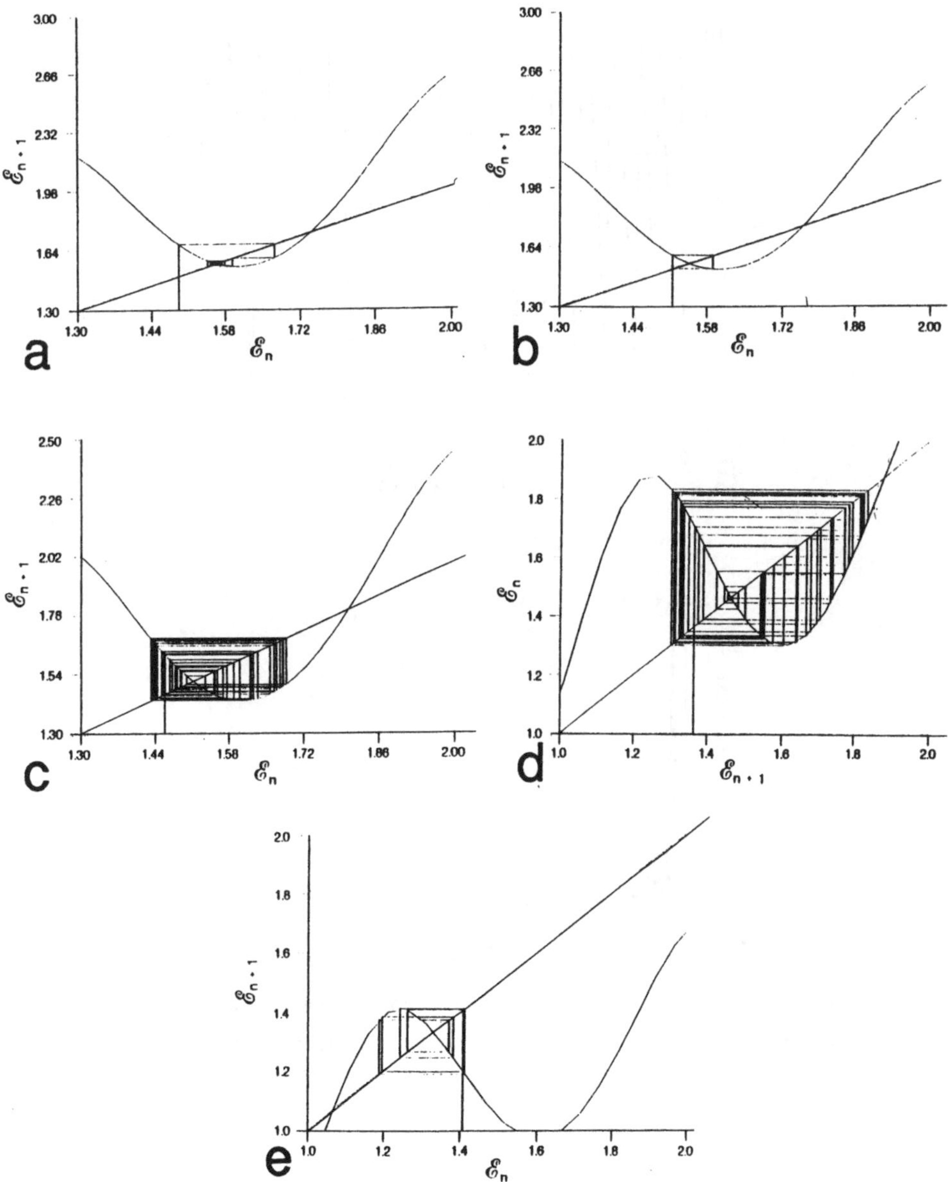

Figure 5. Iterates of the return map $\mathscr{D}$ in the vicinity of the fixed point $\mathscr{S}_3$ for $\tau_{int} = 9.0$. (a) $\alpha = 0.95$ -- the system evolves to the fixed point $\mathscr{S}_3$; (b) $\alpha = 0.93$ -- period-2; (c) $\alpha = 0.88$ -- chaos; (d) $\alpha = 0.8$ -- fully developed chaos; (e) $\alpha = 0.6$ -- period-8.

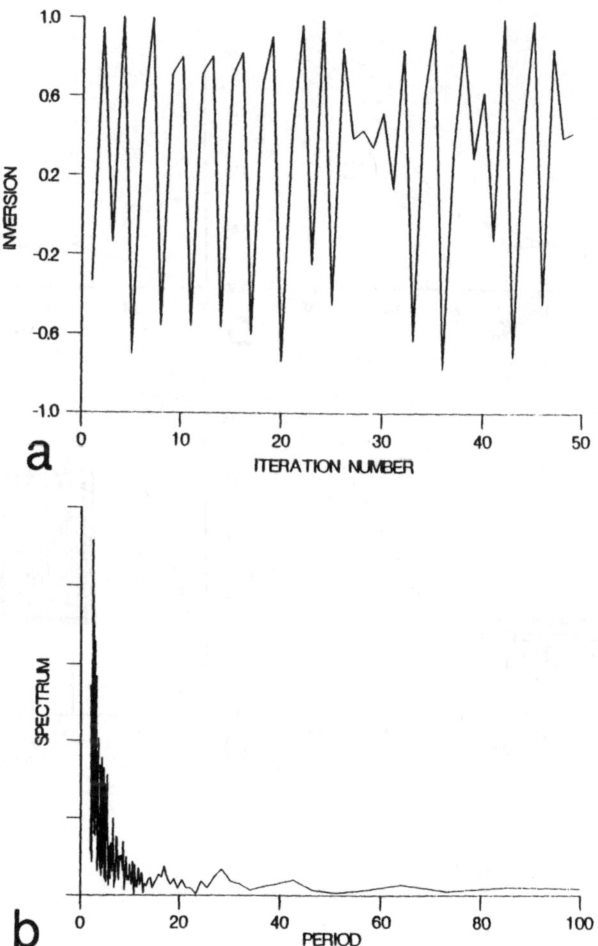

Figure 6. (a) Iterates of the atomic inversion at the exit of the cavity for $\tau_{int} = 9.0$ and $\alpha = 0.80$ and (b) its Fourier transform, exhibiting a broad spectrum.

quantum mechanical interactions. Still, this suggests studying in detail the boundaries between the basins of attraction of the various attractors and determining if it is also a fractal set. An extension of the model to the case where the atoms and cavity mode are detuned is also presently being pursued.

This meeting witnesses the considerable interest in studying quantum systems whose classical version exhibits deterministic chaos. It is hoped that these efforts will help shed light on the elusive "quantum chaos." Considerable work along these lines are being performed in particular by Casati and coworkers (Casati, 1986), Smilanski and coworkers (Blümel et al. 1986), and Graham and coworkers, to mention just a few groups. The micromaser is a truly quantum-mechanical system which is within reach of experimental investigations (Meschede et al., 1985). Because of its simplicity and the exceedingly high degree of control under which experiments can be performed, we feel that it represents an almost ideal alternative system to study quantum chaos. In recent work, we presented a complete quantum theory of the micromaser, including resonator losses (Filipowicz et al., 1986b). We concentrated however on steady-state characteristics and spent very little efforts on its dynamics. In view of the present results, a detailed study of the dynamics of the quantum-mechanical micromaser is called for, and it will hopefully contribute to a better understanding of quantum chaos in dissipative quantum mechanical systems. This study will also lead to a deeper insight into the behavior of lasers and masers at the truly microscopic scale.

ACKNOWLEDGMENTS

We have benefited from numerous discussions on the micromaser with P. Filipowicz and J. Javanainen, as well as with J. H. Eberly, M. O. Scully, and H. Walther. This work is supported in part by the National Science Foundation Grant PHY-8603368 and by the Joint Services Optics Program. PM is grateful to ONR for partial travel support to attend this meeting.

REFERENCES

Allen L. and Eberly J. H., 1975, *Optical Resonance and Two-Level Atoms* (Wiley, New York).

Barnett S. M. and Knight P. L., 1985, Phys. Rev. **A33**, 2444.

Barnett S. M., Filipowicz P., Javanainen J., Knight P., and Meystre P., 1986, in *Frontiers in Quantum Optics*, eds. E. R. Pike and S. Sarkar (Adam Hilger, Bristol and Boston).

Blümel R., Fishman S., and Smilanski U., 1986, J. Chem. Phys. **84**, 2604.

Casati G., 1986, in *Optical Instabilities*, Cambridge Studies in Modern Optics Vol. 4, eds. R. W. Boyd, M. G. Raymer, and L. M. Narducci (Cambridge University Press, Cambridge), and references therein.

Cummings F. W., 1965, Phys. Rev. **140**, 1051.

Eberly J. H., Narozhny N. B., and Sanchez-Mondragon J. J., 1980, Phys. Rev. Lett. **44**, 1323.

Feigenbaum M. J., 1978, J. Stat. Phys. **19**, 25.

Feigenbaum M. J., 1979, J. Stat. Phys. **21**, 669.

Filipowicz P., Javanainen J., and Meystre P., 1986a, Optics Commun. **58**, 327.

Filipowicz P., Javanainen J., and Meystre P., 1986b, Phys. Rev. **A**, in press.

Filipowicz P., Javanainen J., and Meystre P., 1986c, J. Opt. Soc. Am. **B3**, 906.

Gabrielse G. and Demhelt H., 1985, Phys. Rev. Lett. 55, 67.

Graham R., 1986, in *Optical Instabilities*, Cambridge Studies in Modern Optics Vol. 4, eds. R. W. Boyd, M. G. Raymer and L. M. Narducci (Cambridge University Press, Cambridge), and references therein.

Goy P, Raimond J. D., Gross M., and Haroche S., 1983, Phys. Rev. Lett. **50**, 1903.

Gwinn E. G. and Westervelt R. M., 1986, Phys. Rev. **A33**, 4143.

Hulet R. and Kleppner D., 1985, Phys. Rev. Lett. 55, 2137.

Jaynes E. T. and Cummings F. W., 1963, Proc. IEEE 51, 89.

Meschede D., Walther H., and Müller G., 1985, Phys. Rev. Lett. **54**, 551.

Schuster H. G., 1984, *Deterministic Chaos: An Introduction* (Physik Verlag, Weilheim).

Walther H., 1985, Private communication.

CHAOS IN A DRIVEN QUANTUM SPIN SYSTEM

H J MIKESKA AND H FRAHM

1. INTRODUCTION

We present an investigation of the dynamics of one anisotropic spin in an external, periodically time-dependent field. This is a simple quantum system, which displays chaotic behaviour in the classical limit. Chaotic behaviour in classical deterministic systems is characterised by an extremely sensitive dependence on initial conditions; we are interested in the fundamental question, how this classical behaviour translates to the corresponding quantum system.

Previous investigations of this question, mainly using the model of the kicked rotator (Casati et al. 1978, Fishman et al. 1982, Shepalyansky 1983, Hogg and Huberman 1983 and others) have shown that quantum-mechanical effects decrease classical stochasticity. This is generally clear from the fact that Schrödingers equation is linear. The interesting point is the approach of the classical limit. A spin system is particularly appropriate for an investigation of this point owing to its finite Hilbert space. In the following we will discuss stochasticity with the strength of quantum effects, measured by the inverse of the spin length S, as parameter. We will report the essential results of our work and refer to the original publications (Frahm and Mikeska 1985, 1986) for further details. It should also be mentioned that a model which is very similar to our model has been investigated independently by Haake et al. (1986).

In our context we can consider two different aspects of quantum mechanics, which both will be related to our problem:

(i) The aspect of indeterminism, connected intimitately to the wavefunction description and the uncertainty relation. As a consequence of the sensitive dependence on initial conditions in classical chaotic systems their dynamical behaviour is de facto not predictable; a description of the dynamics in terms of trajectories is useless and has to be replaced by a description introducing a probability distribution already on the classical level. When the quantum system can be described by a reasonably localised wavepacket for a sufficiently long time, which defines the semiclassical regime, there will exist two different sources of unpredictability, classical stochasticity and quantum uncertainty, which both are to be described by a probability distribution. We will discuss the relation between these two sources of unpredictability and present a unified description of classical stochasticity and quantum uncertainty in the semiclassical region.

(ii) The aspect of quantisation of energy levels (or quasi-energy levels in the present case of a periodically time-dependent Hamiltonian). Recently considerable activity has focussed on the investigation of the conjecture that regular resp. chaotic classical systems are characterised by different well-defined energy level statistics when considered as a quantum system in its classical limit (see e.g. Bohigas et al. 1984). This leads to the question whether the decrease of stochasticity owing to quantum effects is reflected in a corresponding crossover in their level statistics.

The phase space approach allows a simple qualitative discussion of the importance of quantum effects in a classically chaotic system. Consider a classical Hamiltonian system with two degrees of freedom p, q, and follow the time development of a localised region $\Delta p \Delta q$ in the corresponding two-dimensional phase space. Linearising the deviation from a classical trajectory, two Lyapunov exponents $\lambda_{1,2}$ are obtained, which owing to the Hamiltonian character of the system fulfil $\lambda_1=-\lambda_2=\lambda$. Using an orthogonal transformation $(p,q)\rightarrow(P,Q)$ to orient the axis of the coordinate system parallel to the directions of purely exponential decrease and exponential increase in phase space, the time development of our localised region is written as

$$\Delta_p \Delta_q \to (\Delta_p\ e^{-\lambda t})(\Delta_q\ e^{\lambda t})$$

The quantum-mechanical uncertainty relation introduces a division of phase space in cells of size $\Delta P \Delta Q = \hbar$. Depending on the specific question to be asked this implies uncertainties in the variables P and Q according to $\Delta P \sim \hbar^{\alpha}$, $\Delta Q \sim \hbar^{1-\alpha}$. The classical description becomes inadequate, if the small dimension of the localised region considered above shrinks below the value given from the uncertainty relation, i.e. for (assuming the symmetric case $\alpha = \frac{1}{2}$)

$$\Delta p\ e^{-\lambda t} < \text{const}\ \hbar^{1/2}$$

This leads to a crossover time

$$\tau = \frac{1}{2\lambda}\ \ln\ \frac{\text{const}}{\hbar} \tag{1.1}$$

For $t<\tau$ semiclassical trajectories are meaningful, whereas for $t>\tau$ quantum interference effects dominate classical stochasticity. The different regimes are shown in Fig. 1. The abscissa of this plot is identical with the classical system, whereas

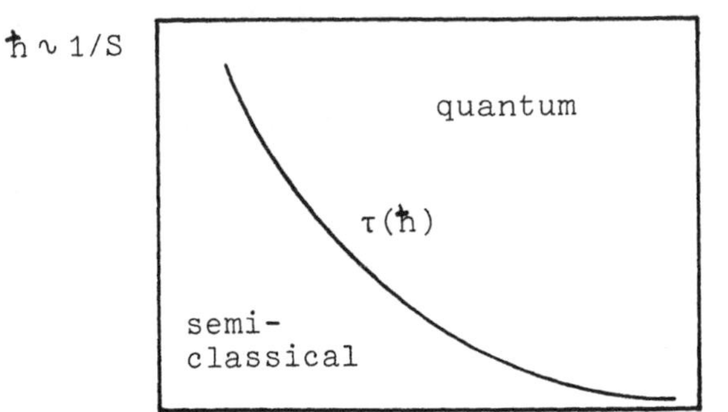

Fig. 1 Different regimes for a classically chaotic quantum system

for any finite strength of quantum effects one always enters the genuine quantum regime. Thus it is clear that the limits $t\to\infty$ (essential for a mathematically strict definition of chaos) and $\hbar \to 0$ do not commute. Our discussion will refer mainly to the semiclassical regime, where the concepts of classical chaotic theory are applicable for some finite time; much less can be said about the genuine quantum regime.

The chapter is organised as follows: We introduce our model in section 2 and present the phase space approach and the level statistics approach to the semiclassical regime in sections 3 and 4 respectively. Further discussions and final conclusions are given in section 5.

2. THE MODEL

Our model is defined by the Hamiltonian

$$H = A(S^z)^2 - \mu B_o \vec{f}(t)\vec{S} \tag{2.1}$$

with spin operators S^α and a periodic magnetic field, $\vec{f}(t+T) = \vec{f}(t)$. The model is a quantum-mechanical generalisation of the corresponding classical planar model discussed by Croquette and Poitou (1981). These authors considered a compass needle in a magnetic field $\vec{B}(t)=B_o\cos\omega t\vec{e}_x$. This classical planar model contains chaotic trajectories for any B_o and displays a transition to complete irregularity at a critical value of the field amplitude B_o. This model can be obtained from (2) by taking the classical limit and assuming large values of A to suppress the z-component of spin.

The parameters of our model are $\mu B_o S$, AS^2, S and $\omega = 2\pi/T$. It is useful to measure time in units of ω^{-1}, i.e. $\omega t \to t$, and to introduce the following dimensionless parameters:

planarity parameter $$b = \frac{\mu B_o}{A\hat{S}} \tag{2.2}$$

stochasticity parameter $$s = \frac{2}{\omega}\sqrt{A\hat{S}\mu B_o} \tag{2.3}$$

quantum parameter $$\hat{S} = \sqrt{S(S+1)} \tag{2.4}$$

2.1 The classical system

Introducing polar angles according to

$$\vec{S} = S\{\cos\Theta\cos\phi,\ \cos\Theta\sin\phi,\ \sin\Theta\} \tag{2.5}$$

our system is defined by the Hamiltonian

$$H = AS^2\sin^2\Theta - \mu B_0\vec{f}(t)\vec{S} \tag{2.6}$$

and the Poisson bracket

$$\{S^z,\phi\} = 1 \tag{2.7}$$

For an oscillating magnetic field $\vec{f}(t) = \vec{e}_x\cos t$ one obtains the equations of motion

$$\frac{d\Theta}{dt} = -\frac{s\sqrt{b}}{2}\cos t\sin\phi \tag{2.8a}$$

$$\frac{d\phi}{dt} = \frac{s}{\sqrt{b}}\sin\Theta + \frac{s\sqrt{b}}{2}\cos t\cos\phi\,\mathrm{tg}\,\Theta \tag{2.8b}$$

In the planar limit, $b<<1$, these reduce to

$$\frac{d^2\phi}{dt^2} = -\frac{s^2}{4}\left[\sin(\phi-t) + \sin(\phi+t)\right] \tag{2.9}$$

(2.9) is identical to the equation studied by Croquette and Poitou (1981) and allows to identify s as stochasticity parameter. We note that our model with two degrees of freedom ($q_1 = \phi$, $p_1 = \sin\Theta$) and a time-dependent external force is equivalent to a system with four degrees of freedom ($q_1, p_1, q_2 \equiv t$, p_2) without explicit time dependence, i.e. a Hamiltonian system. Thus we can use energy conservation and consider Poincaré sections for $q_2 \equiv t = t_n = 2\pi n$ (stroboscopic representation) to obtain a two-dimensional phase space. Trajectories in this phase space for $s \lesssim 1.7$ are either chaotic or regular (with two main fixed points at $\phi = 0$, $\sin\Theta \cong \pm\sqrt{b}/s$), whereas for $s \gtrsim 1.7$ there is one chaotic trajectory, which fills the whole phase space.

Quantitative information about the regularity of the time evolution in classical phase space is obtained by an analysis of the deviations from a classical trajectory $\phi_o(t)$, $\sin\Theta_o(t)$:

$$\frac{d\delta x}{dt} = \frac{s\sqrt{b}}{2} \cos t\{tg\Theta_o \sin\phi_o \delta x - \cos\Theta_o \cos\phi_o \delta y\} \tag{2.10a}$$

$$\frac{d\delta y}{dt} = \frac{s}{\sqrt{b}} \delta x + \frac{s\sqrt{b}}{2} \cos t \{\frac{1}{\cos^3\Theta_o} \cos\phi_o \delta x - tg\Theta_o \sin\phi_o \delta y\} \tag{2.10b}$$

From these equations the Lyapunov exponents $\lambda_{1,2}$ can be found by numerical analysis; owing to the Hamiltonian properties of the system one has $\lambda_1+\lambda_2 = 0$. Irregular trajectories in classical phase space are defined by a positive real part of, say, λ_1.

2.2 The quantum system

A convenient representation of the spin algebra, which is adapted to the symmetry of the system, is given by

$$s^\alpha = S^\alpha/\hbar\hat{S} \tag{2.11}$$

$$s^+ = s^x+is^y = e^{i\phi}\sqrt{1-(s^z)^2-s^z/\hat{S}} \tag{2.12}$$

In this representation our system is defined by the Hamiltonian

$$H = \frac{s\hat{S}}{2\sqrt{b}} (s^z)^2 - \frac{s\sqrt{b}\hat{S}}{2} \vec{f}\cdot\vec{s} \tag{2.13}$$

and the commutator

$$[\phi,s^z] = i/\hat{S} \tag{2.14}$$

From (2.14) it is clear that in the dimensionless representation (s^z,ϕ) the role of $\hbar$ is played by $\hat{S}^{-1}$ whereas ϕ and s^z are very similar to position and momentum operators (apart from the finite number of dimensions of the spin Hilbert space as reflected in the square root in (2.12)).

For $\vec{f}\sim\vec{e}_x$ Schrödinger's equation in s^z representation

$$|\psi\rangle = \sum_m c_m(t)|m\rangle \tag{2.15}$$

is given by

$$\frac{d\,c_m}{dt} = -i\hat{S}\,\frac{s}{2\sqrt{b}}\{\frac{m^2}{\hat{S}^2}c_m - \frac{b}{2}f(t)\sqrt{1-\frac{m(m-1)}{\hat{S}^2}}\,c_{m-1} - \frac{b}{2}f(t)\sqrt{1-\frac{m(m+1)}{\hat{S}^2}}\,c_{m+1}\} \tag{2.16}$$

The formal solution is found using Floquet's theorem because of the periodicity $f(t+2\pi) = f(t)$

$$c_m^{(\alpha)}(t) = e^{-i\mu_\alpha t}\,f_m^{(\alpha)}(t) \tag{2.17}$$

with

$$f_m^{(\alpha)}(t+2\pi) = f_m^{(\alpha)}(t)$$

Thus the time development of the system is quasiperiodic and chaotic behaviour, strictly speaking, does not exist, as should be clear from the linearity of Schrödinger's equation. In the classical limit $S\to\infty$, however, the number of Floquet frequencies diverges; the subject of the field generally called quantum chaos is to a large degree to understand, how the classically chaotic behaviour emerges from the quasiperiodic quantum behaviour. Therefore it is of particular importance to study the semiclassical regime, which is the subject of the following sections.

The Floquet frequencies μ_α play the role of quasienergies as can be seen by describing the dynamics by the unitary time evolution operator

$$U(t,t') = T\exp\{-i\int_t^{t'} H(\tau)d\tau\} \tag{2.18}$$

where T is the time ordering operator. Owing to the periodicity of f it is sufficient to consider $U(0,2\pi) =: U$. The eigenvalues e_α of U are related to the Floquet frequencies by

$$e_\alpha = e^{2\pi i\mu_\alpha} \tag{2.19}$$

U may be written as a direct product of operators U^+ and U^-, acting on states with even, resp. odd parity, since the parity $\exp(i\pi S^x)$ is a conserved quantity.

3. SEMICLASSICAL REGIME: PHASE SPACE APPROACH

We want to describe our system in the semiclassical regime in a way to include the classical phase space description in the limit $S\to\infty$. For this purpose we consider the wave function in the S^z representation (2.15) and assume that the amplitudes c_m can be parametrised according to

$$c_m \sim \exp \hat{S}\left[-i\phi_o \frac{m-m_o}{\hat{S}} - \frac{\Gamma}{2}\frac{(m-m_o)^2}{\hat{S}^2}\right] \tag{3.1}$$

This ansatz corresponds to a WKB approximation and is motivated by the fact that to $O(1/S)$ and with $\Gamma = (1-\frac{m_o^2}{\hat{S}^2})^{-1}$ it is the solution to the eigenvalue equation $\hat{S}_{op}|\psi\rangle = \vec{S}|\psi\rangle$ with vector spin operator $\hat{S}_{op}$ and $\hat{S}$ given by (2.5) with $\phi = \phi_o$, $\sin\Theta = \sin\Theta_o = m_o/\hat{S}$. Thus the quantum-mechanical wave function is parametrised by the classical quantities $\phi_o(t)$, $\sin\Theta_o(t)$ and a complex width parameter $\Gamma(t)$. The time dependence of these quantities can be calculated consistently if only terms of lowest order in 1/S are taken into account.

The wavefunction $|\psi\rangle = \sum_m c_m|m\rangle$ can be interpreted in terms of a quasiclassical distribution. According to Wigner (Hillery et al. 1984) a probability distribution P(p,q), which has all the properties of a phase space distribution except positive definiteness, is obtained from a wavefunction $\psi(q)$ via

$$P(p,q) = \frac{1}{\pi\hbar}\int dy\psi^*(q+y)\psi(q-y)e^{2ipy/\hbar} \tag{3.2}$$

For the present wavefunction this leads to ($x=\sin\Theta$)

$$P(x,\phi) = \exp - \frac{\hat{S}}{\Gamma_R} \left[|\Gamma|^2(x-x_o)^2+(\phi-\phi_o)^2-2\Gamma_I(x-x_o)(\phi-\phi_o) \right] \tag{3.3}$$

For our WKB like ansatz $P(x,\phi)$ is even positive definite. In fact, for large values of S our ansatz (3.1) is equivalent to the coherent state formulation for the spin system (Arecchi et al. 1972).

It is instructive to consider the mean square fluctuations, which are obtained by integrating over ϕ, resp. x (or directly from (3.1), resp. its "Fourier transform"):

$$\Delta_x{}^2 = <(\Delta\sin\Theta)^2> = \frac{1}{\hat{S}\Gamma_R} \tag{3.4a}$$

$$\Delta_\phi{}^2 = <(\Delta\phi)^2> = \frac{1}{\hat{S}\mathrm{Re}\ 1/\Gamma} = \frac{1}{\hat{S}} \frac{|\Gamma|^2}{\Gamma_R} \tag{3.4b}$$

The shape of the probability distribution in phase space is indicated in Fig. 2. The major axis of the distribution are

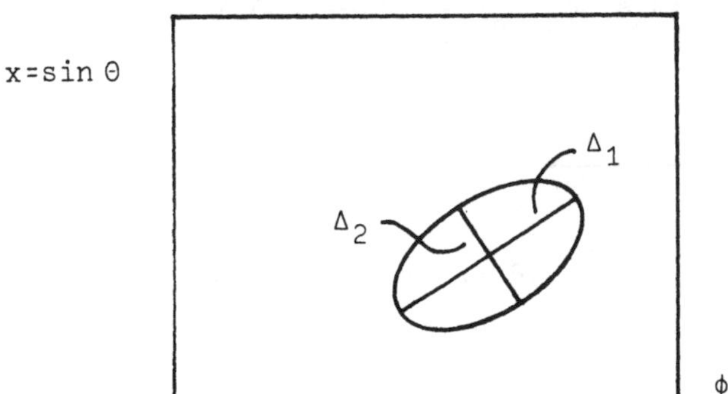

Fig. 2 Probability distribution in phase space

calculated by diagonalising the quadratic form in (3.3):

$$\Delta_{1,2} = \{ \frac{\hat{S}}{2\Gamma_R} \left[1+|\Gamma|^2 \pm \sqrt{(1+|\Gamma|^2)^2-4\Gamma_R{}^2} \right] \}^{-1/2} \tag{3.5}$$

Thus by calculating $\Gamma(t)$ we obtain the shape of a semiclassical distribution function. Analogous to the classical case the existence of a quantum Lyapunov exponent

$$\lambda = \lim \frac{\ln \Delta_1(t)}{t} \tag{3.6}$$

can be investigated. We note $\Delta_1\Delta_2 = 1/\hat{S}$ = const., i.e., independent of time, which corresponds to area conservation for our quantum Hamiltonian system.

For an evaluation of λ we have set up the equations of motion for $\phi_o(t)$, $\sin\Theta_o(t)$ and $\Gamma(t)$ including terms of $O(1/\hat{S})$. It is instructive to use these equations to obtain the equations of motion for the mean squared fluctuations Δ_x, Δ_ϕ:

$$\frac{d\Delta_x}{dt} = \frac{s\sqrt{b}}{2} \cos t\{\mathrm{tg}\Theta_o \sin\phi_o - \frac{q(\Delta)}{\Delta_x^2} \cos\Theta_o \cos\phi_o\}\Delta_x + O(1/\hat{S}) \tag{3.7a}$$

$$\frac{d\Delta_\phi}{dt} = \frac{s\sqrt{b}}{2} \cos t\{\frac{q(\Delta)}{\Delta_\phi^2} \frac{1}{\cos^3\Theta_o} \cos\phi_o - \mathrm{tg}\Theta_o \sin\phi_o\}\Delta_\phi + \frac{s}{\sqrt{b}} \frac{q(\Delta)}{\Delta_\phi} + O(1/\hat{S}) \tag{3.7b}$$

with

$$q(\Delta) = \sqrt{\Delta_x^2\Delta_\phi^2 - 1/\hat{S}^2} \tag{3.8}$$

Terms of first order in $1/\hat{S}$ in (3.7) are not neglected when we evaluate these equations, they are just too long to be reproduced here.

We have $q(\Delta) \geqq 0$ owing to the uncertainty relation. When we neglect all terms of $O(1/\hat{S})$, eqs. (3.7) become linear in Δ_x, Δ_ϕ and are in fact identical to eqs. (2.10) for the classical system. Thus we have obtained a natural extension of the classical formulation of stochastic dynamics to include quantum effects. Eqs. (3.7) form a unified description of classical stochasticity and quantum uncertainty. It is interesting to note that the information about the stability of classical trajectories is contained in the width parameter Γ, which appears as a typical quantum variable. Γ, however, has a well-defined limit for $\hat{S}\to\infty$, and the approach to the classical limit occurs only owing to the extra factor $\hat{S}$ in the exponent of (3.1).

Evaluating eqs. (3.7) numerically the following results are obtained:

i) For rotational trajectories in regular regions of phase space we find

$$\Delta_1(t) \sim \sqrt{\frac{1+\text{const } t^2}{\hat{S}}} \tag{3.9}$$

corresponding to

$$\Gamma(t) = \Gamma(0) + i\gamma t \tag{3.10}$$

This is equivalent to the well-known quantum uncertainty owing to the broadening of a Gaussian wave packet under the influence of dispersion.

ii) For trajectories in stochastic regions of phase space a behaviour

$$\Delta_1(t) \sim \frac{1}{\sqrt{\hat{S}}}\, e^{\lambda t} \tag{3.11}$$

is found for sufficiently short times $t<\tau \sim \ln\hat{S}$. Here the limiting time τ is determined by the validity of our expansion in $1/\hat{S}$. Thus it is the time, where the semiclassical approximation breaks down and can therefore be determined from the condition that our wavepacket is well defined, i.e. $\Delta_x{}^2 \ll 1$, $\Delta_\phi{}^2 \ll 1$. Using (3.4) this leads to

$$1/\hat{S} \ll \Gamma_R \ll \hat{S} \tag{3.12a}$$

$$\Gamma_I \ll \hat{S} \tag{3.12b}$$

From (3.4a) we have $\Gamma_R(t) \sim e^{-2\lambda t}$ for $t<\tau$; using $\Gamma_R(\tau) = 1/\hat{S}$ to determine the limiting time we obtain

$$\tau \cong \frac{1}{2\text{Re}\lambda} \ln \hat{S} \tag{3.13}$$

in agreement with the qualitative estimate (1.1).

Since the limit $t \to \infty$ cannot be performed for a finite strength of quantum effects, the quantum Lyapunov exponent can be calculated approximately only by considering the behaviour at finite, though large time $t \lesssim \tau$. In the present approach, because of $\tau \sim \ln \hat{S}$, this requires extremely large spin values, $S \gtrsim 10^8$. Even for these large spin values, however, the quantum Lyapunov exponent differs from its classical limit. As is shown in Fig. 3, $\lambda(S)$ at finite values of S is smaller than in the

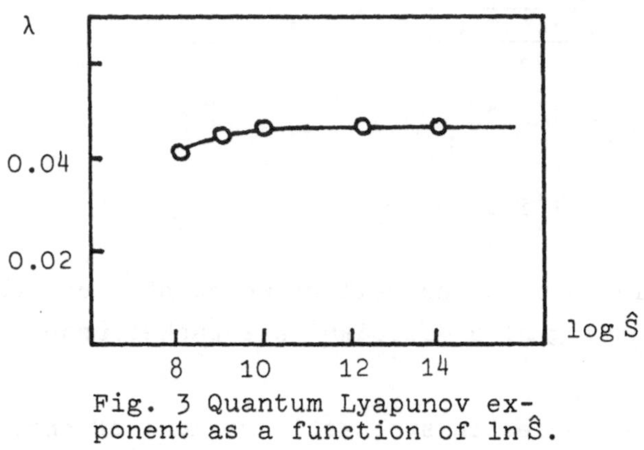

Fig. 3 Quantum Lyapunov exponent as a function of $\ln \hat{S}$.

limit $S \to \infty$. This indicates the decrease of stochasticity owing to quantum effects. Thus our detailed investigation of the phase space approach to our specific model illustrates and supports quantitatively the more general discussion given in the introduction.

The extremely large spin values discussed above are required only for a determination of the Lyapunov exponent, which needs a sufficiently long period of time to be followed. If the time development of a spin component is followed more directly, much smaller values of S, e.g. S = 100, are sufficient to show semiclassical behaviour as can be seen in the work of Haake et al. (1986).

4. SEMICLASSICAL REGIME: LEVELSTATISTICS APPROACH

In this section we discuss the statistics of quasienergy levels μ_α (see 2.17/2.19) for our driven spin system with a twofold purpose: We show that for a periodically driven system the

the statistics of quasienergy levels is equivalent to the energy level statistics in a time-independent system (for a review see Bohigas and Giannoni 1984) and we investigate the dependence of the quasienergy level statistics on the quantum parameter S.

For practical reasons we apply a kicked version of our system

$$\vec{f}(t) = \vec{e}_x \sum_k \delta(t-2\pi k) \tag{4.1}$$

This choice makes it possible to write down explicitly the time evolution operator U (see eq. 2.18)

$$U = e^{-i \frac{\pi S}{\hat{S}\sqrt{b}} (S^z)^2} e^{i \frac{S\sqrt{b}}{2} S^x} \tag{4.2}$$

We have calculated the eigenvalues of U on the computer up to a maximum value S = 400.

It has been conjectured that the statistical properties of the spectrum of a quantum problem in its classical limit are associated with the regularity properties of the corresponding classical system. In particular it has been shown by Berry and Tabor (1977) that the energy levels of classically integrable systems obey Poisson statistics whereas classically chaotic systems are supposed to be equivalent to systems discussed in random matrix theory (Mehta 1967), in particular the Gaussian Orthogonal Ensemble (GOE). The specific quantities considered are the distribution of the spacings between adjacent energy levels P(x) (x is the level spacing measured in units of the mean level spacing) and the rigidity

$$\Delta_3(L) = \frac{1}{L} \min_{A,B} \int_x^{x+L} d\varepsilon \left[n(\varepsilon)-A\varepsilon-B\right]^2 \tag{4.3}$$

where $n(\varepsilon)$ is the number of states with energy below ε. For Poisson statistics one has

$$P(x) = e^{-x} \tag{4.4a}$$

$$\Delta_3(L) = \frac{L}{15} \tag{4.4b}$$

whereas for GOE

$$P(x) \cong \frac{\pi}{2} x e^{-\frac{\pi}{4} x^2} \qquad (4.5a)$$

$$\Delta_3(x) \cong \frac{1}{\pi^2} \ln L - 0.007 \qquad (L \geq 15) \qquad (4.5b)$$

(4.5a) is the Wigner-surmise which reproduces very well the more complicated rigorous result for GOE (Mehta 1967). A transition from Poisson- to GOE-statistics should be found for a system whose classical version contains a parameter that controls its nonintegrability, as the stochasticity parameter s does in our model. Transitions fo this type have been discussed for anharmonic oscillator models by Haller et al. (1984) and Seligman et al. (1985). The transition may be defined quantitatively by an additional parameter ρ in the statistical distributions (Pandey 1979, Berry and Robnik 1984). ρ is the relative volume of the chaotic part of the phase space, i.e. $\rho = 0$ for a classically regular and $\rho = 1$ for a classically chaotic system.

We have analysed the spectra of U numerically using these concepts. The quasienergies are considered in the basic interval $0<\mu_\alpha\leq 2\pi$. Owing to the periodicity of the μ_α the rigidity is meaningful only for L<N, where N=S is the total number of energy levels. Since, however, the asymptotic behaviour becomes apparent for L≤30, the investigation of the rigidity is meaningful for N>100 (allowing the necessary average over x, based on the assumption of a homogeneous distribution of quasienergies).

Our numerical results have been obtained for the value b= 0.077 of the planarity parameter. The distribution of level spacings for S=400 shows a nice transition from Poisson to GOE statistics, and calculations for smaller values of S (200,100) show some evidence for a tendency towards Poisson-like statistics for the pure quantum case of small S. Statistically more significant are the results for the rigidity Δ_3. In Fig. 4 we show the variation of $\Delta_3(L)$ with S for s = 1.6. Whereas for S=400 the GOE distribution characterizing the classical system is practically reproduced, there is a significant shift

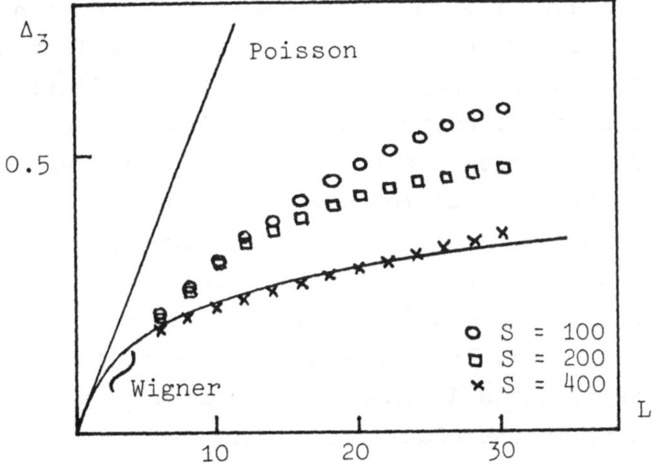

Fig. 4 $\Delta_3(L)$ for s=1.6 and various S

towards a Poisson distribution for the spin values 200 and 100. An analysis in terms of the parameter ρ introduced above leads to the results shown in Fig. 5. The transition from regular (ρ=0) to chaotic (ρ=1) behaviour with increasing stochasticity

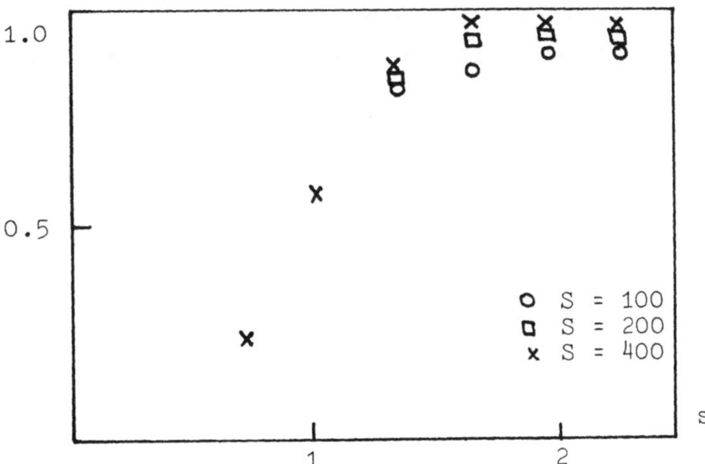

Fig. 5 Transition from regular to chaotic behaviour with stochasticity parameter s for various S

parameter s occurs for all spin values. The level statistics parameter ρ clearly decreases with decreasing S, showing the tendency to more regular behaviour when the strength of quantum effects is increased.

Although the values of S investigated in this section are much smaller than those of the previous section, the fundamental limitations for the two approaches are very similar: In the level statistics approach the number of levels has to be sufficiently large to allow a statistical analysis, which requires $S \gg 1$. Rigorous statistical results can be obtained for $S \to \infty$ only, i.e. in the classical limit. This corresponds to the statement in section 3 that rigorous results on the stochasticity of a system discussed in phase space require the limit $t \to \infty$, i.e. again $S \to \infty$. Nevertheless it appears that quantum effects can be isolated in the semiclassical regime, using the same concepts as in the classical limit.

It is of course tempting to try to connect the two different approaches to quantum properties of classically chaotic systems discussed so far, the phase space approach and the level statistics approach, by considering the dynamics in the context of random matrix theory. The simplest approach to this question is to start with a state $|\psi\rangle = \sum_m \alpha_m |m\rangle$ and to follow its time development, i.e. to calculate $\alpha_m(t)$, with the purpose to discuss the physically relevant quantity $\overline{|\alpha_m(t)|^2}$ (the average is over the appropriate random matrix ensemble). Denoting an energy eigenstate of a particular member of the ensemble by $|i\rangle$, $H|i\rangle = E_i|i\rangle$, we have

$$|i\rangle = \sum_m U_{mi} |m\rangle \tag{4.6}$$

$$\alpha_m(t) = \sum_{i,n} \alpha_n \, U_{ni}^* \, U_{mi} \, e^{-iE_i t} \tag{4.7}$$

$$|\alpha_m(t)|^2 = \sum_{i,j} e^{-i(E_i-E_j)t} f_{m,ij} \tag{4.8a}$$

with

$$f_{m,ij} = U_{mi} \, U_{mj}^* \sum_{n,n'} \alpha_n \alpha_{n'}^* \, U_{ni}^* \, U_{n'j} \tag{4.8b}$$

Performing the random matrix ensemble average for $|\alpha_m(t)|^2$, the average of the product in (4.8a) factorizes (Mehta 1967). Owing to the invariance properties of Gaussian ensembles $\overline{f_{m,ij}}$ is given by (N is the dimension of Hilbert space)

$$\overline{f_{m,ij}} = \frac{1}{N^2} d_m + \frac{1}{N}(c_m - \frac{1}{N} d_m) \delta_{ij} \quad (4.9)$$

i.e. all diagonal and all off-diagonal elements are equal for a given value of m. Thus the time dependence of $\overline{|\alpha_m(t)|^2}$ is determined by the average of $\sum_{ij} \exp(E_i-E_j)t$; this is given completely by the Fouriertransform of the two-level correlation function $\gamma_2(r)$:

$$\sum_{ij} \overline{e^{i(E_i-E_j)t}} = N^2 \int dr\ e^{irt} \gamma_2(r) = :N^2 g(t) \quad (4.10)$$

Using (4.9), one obtains

$$\overline{|\alpha_m(t)|^2} = |\alpha_m(t=0)|^2 + d_m(g(t)-1) \quad (4.11)$$

For Gaussian ensembles in the limit $N \to \infty$ g(t) has been calculated explicitly (Mehta 1967); for GOE it is given by

$$\begin{aligned} g(t) &= 1-2t+t \ln (1+2t) \qquad t \leqq 1 \\ &= -1 + t \ln 2t+1/2t-1 \quad t \geqq 1 \end{aligned} \quad (4.12)$$

where time is measured in units of 2π times the reciprocal of the average level spacing. For the Poisson distribution characterising regular systems γ_2 is constant apart from the self-correlation term $\delta(r)$, i.e. g(t)=1 corresponding to a constant value of $\overline{|\alpha_m(t)|^2}$. For GOE, $\overline{|\alpha_m(t)|^2}$ approaches a new equilibrium value for large time; the approach to equilibrium is proportional to t^{-2}. Similar results have recently been published for correlation functions in GUE (Casati et al. 1986).

However, the large time behaviour obtained in this way to our opinion cannot be related to the time dependence in phase space as discussed in the phase space approach: In the random matrix approach the correlations between eigenvalues and eigenvectors present for one particular system are lost, leading to a smoothing of dynamical correlations. The characteristic time and the asymptotic behaviour of g(t) does not relate to stochastic dynamics but merely characterises the decay of fluctuations owing to the presence of different members of the random matrix ensemble. Thus it appears that random matrix theory is not suitable to describe the dynamics of stochastic systems in any direct way.

5. DISCUSSION AND CONCLUSIONS

We have investigated a simple driven quantum spin system, which is chaotic in the classical limit. Using the complementary approaches of phase space analysis and level statistics we have shown in both schemes that for large spins, i.e. in the semiclassical regime, stochasticity can be discussed using concepts developped for classical systems; quantum effects show up in quantitative changes of the Lyapunov exponent and the level distribution function. Although the distribution of quasienergy levels agrees with the one obtained from random matrix theory, this relation appears not to be true generally since random matrix theory does not reproduce the dynamics of the system.

A detailed description of our system can be given explicitly only in the semiclassical region; a theory covering completely the transition from the classical to the full quantum limit is not available so far. A possibility to study the regime of smaller spin values may open up from recent work (Nakamura et al. 1986) on the same system as was considered here: These authors investigated in more detail the quasienergy eigenfunctions in terms of coherent states. They find that these eigenfunctions are characterised by a fractal structure. The fractal dimension equals 1 in the regular region and increases above 1 when the stochsticity parameter increases. Although the determination of the fractal dimension also requires a sufficiently fine division of phase space, i.e. large spins, a study of wavefunctions instead of wavepackets or eigenvalues is probably suited better for an extension to the full quantum regime.

Finally we want to discuss the relation of our model to that of the much studied kicked rotator. The most obvious difference between these two models is that the operator $s^z = S^z/\hat{S}$ has a discrete spectrum limited to the region -1 to +1, whereas angular momentum as one of the variables in the kicked rotator problem has a continuous unbound spectrum. Thus energy diffusion, which is the most direct indicator of classical chaos for the kicked rotator, cannot be considered in the spin model with its upper bound for the value of the energy. On the other hand the finite Hilbert space of the spin model greatly facilitates the investigation of the approach of the classical limit, in particular in the levelstatistics approach. Actually it has been shown that the kicked quantum rotator can be rewritten as an Anderson model (Fishman et al. 1982) and thus has a Poisson-like quasienergy level distribution (Molcanov 1981). It is possible, however, that this result is a peculiarity of the kicked rotator and not generically representative for periodically driven quantum systems. Thus the two models are similar but not identical and allow to approach common questions under different aspects. We expect that further investigation of both models in the future will contribute to an understanding of the open questions, in particular in the fully quantummechanical regime.

References

Arecchi, F.T., Courtens, E., Gilmore, R. and Thomas, H. 1972, Phys. Rev. A6 2211
Berry, M.V. and Tabor, M 1977, Proc. Roy. London A356 375
Berry, M.V. and Robnik, M. 1984, J. Phys. A17 2413
Bohigas, O., Giannoni, M.J. and Schmit, C. 1984, Phys. Rev. Lett. 52 1
Bohigas, O. and Giannoni, M.J. 1984, Lecture Notes in Physics Vol. 209 (Springer, Berlin) 1
Casati, G., Chirikov, B.V., Israilev, F.M. and Ford, J. 1978, Lecture Notes in Physics Vol. 93 (Springer, Berlin) 334
Casati, G., Guarneri, I. and Mantica, G. 1986, Physica 21D 105
Croquette, V. and Poitou, C. 1981, J. Phys. (Paris) 42 L537
Fishman, S., Grempel, D.R. and Prange, R.E. 1982, Phys. Rev. Lett. 49 509
Frahm, H. and Mikeska, H.J. 1985, Z. Phys. B60 117
Frahm, H. and Mikeska 1986, Z. Phys. B, in print
Haake, F., Kus, M., Mostowski, J. and Scharf, R. 1985, preprint, Univ. Essen
Haller, E., Köppel, H. and Cederbaum, L.S. 1984, Phys. Rev. Lett. 52 1665
Hillery, M., O'Connell, R.F., Scully, M.O. and Wigner, E.P. 1984, Phys. Rep. 106 121

Hogg, T. and Huberman, B.A. 1983, Phys. Rev. A28 22
Mehta, M.L. 1967, Random Matrices and the Statistical Theory of Energy Levels, Academic Press (New York)
Molcanov, S.A. 1981, Comm. Math. Phys. 78 429
Nakamura, K., Okazaki, Y. and Bishop, A.R. 1986, Phys. Rev. Lett. 57 5
Pandey, A. 1979, Ann. Phys. (N.Y.) 119 170
Seligman, T.H., Verbaarschot, J.J.M. and Zirnbauer, M.R. 1985, J. Phys. A18 2751
Shepelyansky, D.L. 1983, Physica D8 208

FIXED POINTS AND CHAOTIC DYNAMICS OF AN INFINITE DIMENSIONAL MAP

J V MOLONEY, H ADACHIHARA, D W McLAUGHLIN AND A C NEWELL

1. INTRODUCTION

An appealing idea of modern dynamics is that the complicated and apparently stochastic time behaviour of large and even infinite-dimensional nonlinear systems is in fact a manifestation of a deterministic flow on a low dimensional chaotic attractor. If the system is indeed low-dimensional, it is natural to ask whether one can identify the physical characteristics such as the spatial structure of those few active modes which dominate the dynamics. Our thesis is that these modes are closely related to and best described in terms of asymptotically robust, multiparameter solutions of the nonlinear governing equations. In this article we present a review of our progress to date in isolating such nonlinear spatial modes and identify their role in inducing chaotic dynamics. Our mathematical model is an infinite dimensional nonlinear map derived from a well known model in nonlinear optics. We identify a type of chaos which is best described as a phase or weak turbulence and which arises when there is an endless competition between equally resilient, spatially localised coherent structures which are infinite time asymptotic states and which are initiated at random parts of the physical domain. The coherent spatial structures in our problem are the solitary wave solutions of nonlinear partial differential equations which belong to a well known class of integrable or near integrable systems, namely, nonlinear Schrodinger type equations.

There are a number of other approaches to the study of chaos in nonlinear physical systems. Indeed, one of the great successes of modern nonlinear dynamics has been the discovery of universal scenarios, which are associated with iterates of rather simple looking mathematically abstract maps such as the logistic or circle map, and which mimic, in some instances, the transition from a simple to a chaotic state in relatively complex physical systems. Amongst the best known transitions to chaos are period doubling, the Ruelle-Takens-Newhouse route via quasiperiodicity and intermittency (Shuster, 1984). A rather unsatisfactory aspect of such abstract maps is that the bifurcation parameter appearing in them has no obvious relation to any physical parameter. Real physical systems, being intrinsically infinite dimensional, are more likely to have multiple attractors with their individual basins of attraction spanning the infinite dimensional phase space. Different attractors may only be accessible by varying one or more, of a large set of available physical parameters.

A somewhat different theoretical approach is to attempt to reduce a complicated physical model to a problem involving just a few physically relevant modes. In some instances the mode projection and truncation is dictated by the geometry of the physical problem; a classical example is the derivation of the Lorenz equations which describe the behaviour of convection roles in Rayleigh-Benard fluid convection. These equations lose their validity under larger physical stresses where increasingly strong nonlinear interactions become evident. Extension of the problem to include progressively more modes leads to the unsatisfactory situation that the observed routes to chaos become sensitive to the level of truncation.

The main thrust of our study is to combine full blooded numerics on the infinite dimensional problem with mathematical analysis, the latter, in some instances, being motivated by our numerical observations. Using this mode of attack, we find that while solitary waves represent the dominant coherent spatial nonlinear modes in the problem, we cannot entirely ignore the effects of radiation modes (quasi-plane wave solutions), even though our choice of physical problem parameters ensures that they represent small perturbations of the former. In fact these radiation modes play a central role in inducing chaotic dynamics by offsetting the stabilising tendencies of the solitary waves. Our analysis later in this chapter will lead us to conclude that a new type of modulational chaos is well described by a set of nonlinear basis functions consisting of solitary waves and nonlinear radiation modes.

Our physical model, derived from nonlinear optics, consists of a passive externally pumped ring resonator as sketched in Figure 1. This model has been extensively treated in the optics literature (Bowden et al., 1984, Gibbs et al., 1986, Gibbs, 1985) and among some of its potentially useful physical properties is its ability to function as an optical memory. Hysteresis phenomena will play a central role in inducing much of the interesting nonlinear dynamical behaviour of the system. We emphasise that our conclusions regarding the dynamical behaviour of this nonlinear optical system should have a much wider impact beyond this specific model. The optics model discussed here also highlights the difficulties associated with inapproriate truncation for the purpose of simplification. While it may seem appropriate to adopt a plane wave approximation we find that much of the relevant physics is ignored by so doing. The plane wave optics map, however, does remain a mathematically interesting object and we shall find it useful in later sections in orienting our study of the more complex infinite dimensional problem.

The main results to be presented in the later sections are summarised now for convenience. After developing the model of the infinite dimensional map with some accompanying physical motivation in section 2 we proceed in section 3 to determine its fixed points numerically in function space. At this stage we specialise to a one transverse dimensional problem. The nature of the fixed points is shown to be dependent on the form of nonlinearity used in the model. Specifically, we consider both saturable and cubic (Kerr) nonlinearity. In section 4 we motivate the soliton projection technique which allows us to reduce the infinite dimensional map to a two dimensional map in the soliton parameters (amplitude and phase). The cases of cubic (Kerr) and saturable nonlinearities are considered separately. The fixed points of the reduced maps are compared to the "exact" numerically generated fixed points from section 3. We find excellent agreement for the saturable case with less spectacular but satisfactory agreement in the Kerr case. In section 5 we numerically study the effect of larger stress on the fixed points of the infinite dimensional map. The parameter region of interest lies close to the initial period doubling bifurcation of the plane wave optics map. A broad shelf, which we associate in section 3 with a lower (bistable) branch quasi-plane wave fixed point, assumes an important role in initiating the instability. The significant observation at this stage is that the instability bears no resemblance to that observed in the plane wave map and is seeded by transverse fluctuation. Section 6 is devoted to the analysis and explanation of this phenomenon as a new type of modulation instability. In this section we go to some pains to understand the role of spatial sideband growth in causing the instability and successfully identify the participating spatial modes.

2. OVERVIEW OF THE THEORY

The mathematical model describing the propagation of a laser beam in an optical ring cavity has been extensively discussed in the literature usually within the plane wave approximation. For the specific case of the infinite dimensional problem to be discussed here the reader is referred to the papers by Aceves et al. (1986), McLaughlin et al. (1983). We shall be content here with a brief sketch of the mathematical equations. The nonlinear partial differential equation (PDE) describing beam propagation through the nonlinear medium within the cavity, is given by

$$2i \frac{\partial G_n}{\partial z} + \gamma \nabla_T^2 G_n + pN(G_n G_n^*)G_n = 0 \qquad (1)$$

where $G_n(\underline{x},z)$ represents the complex envelope of the electromagnetic field which will generally be assumed finite in the transverse $\underline{x} \equiv (x,y)$ direction and propagates in the z-direction. The index n represents a discrete time (map index) which counts the number of circuits of the beam around the resonator (Figure 1). The coefficient γ of the two dimensional transverse Laplacian $\nabla_T^2 = \partial^2/\partial x^2 + \partial^2/\partial y^2$ measures the importance of diffraction (transverse diffusion) of the beam. This coefficient γ is defined in terms of the more familiar Fresnel number F as follows, $\gamma = \ell n2/4\pi F$.

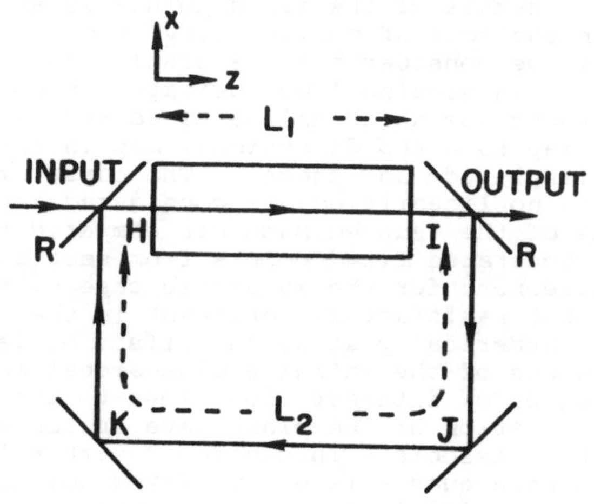

Figure 1. Schematic diagram of an optical ring resonator. The laser beam incident from the left passes through a partially transmitting mirror of intensity reflectivity R; the mirror transmission T = 1-R. Next the beam propagates through a cell of length L_1 containing a medium with a nonlinear response to the applied optical field. Part of the beam is monitored through the output mirror and the remainder is fed back on a return circuit of length L_2 to add on to the injected pump beam. We assume, for the purpose of the present article that the resonator is filled with the nonlinear medium, i.e. $L_2 = 0$.

Therefore increasing the Fresnel number, which corresponds to increasing laser focal spot size for fixed wavelength, weakens the diffractive coupling. The final term in eqn. (1) is the nonlinear term and for the purpose of the present article may assume one of two forms; $N(I^*) = -1/1+I$ $(I = GG^*)$ a saturable nonlinearity or $N(I) = -1 + 2I$ a Kerr nonlinearity. Notice that the Kerr nonlinearity may be viewed as the first order low intensity $[|G_n|^2 \ll 1]$ expansion of the saturable term. The coefficeint p multiplying the nonlinear term is a measure of the effective nonlinear medium propagation length and is defined in terms of the fundamental material constants and physical medium length L_1. In the present article we shall assume that the nonlinear medium fills the ring cavity so that $L_2 = 0$ in Figure 1. The larger the coefficient p the greater the influence of the nonlinearity; it is important to realise that the physical propagation length L_1 itself need not be large to achieve large p.

It is a straightforward matter to show that a plane wave satisfies the full nonlinear partial differential equation (PDE) (eqn. 1) by direct substitution. A cautionary note however. While a plane wave may satisfy eqn. (1), it may not represent a nonlinearly stable solution. Section 6 is devoted to the analysis of this situation,. In addition equation (1) has another class of particularly interesting fully nonlinear solutions when $p > 0$ (self-focussing). These latter solitary wave solutions arise when one seeks solutions to equation (1) subject to the boundary conditions $G_n(\underline{x},z) \to 0$ as $|\underline{x}| \to \infty$; in other words these are transversely confined nonlinear waves. They exist due to a sensitive balance between linear diffractive spreading and nonlinear self-focussing. These solitary waves have the important property of being extremely robust, retaining their forms under the action of various types of perturbation. In the special case of the optical Kerr nonlinearity, $N(I) = -1+2I$, equation (1), in one transverse spatial dimension, is precisely the well known nonlinear Schrodinger equation, one of a special class of exactly integrable PDE's. This latter equation admits N-soliton waves as solutions which possess a number of remarkable properties (Scott et al., 1973, Newell, 1984).

The mathematical problem describing beam propagation in the ring resonator of Figure 1 is not fully posed until we specify the boundary conditions. In appropriately scaled coordinates (McLaughlin et al. (1983,1985) Moloney (1986a), Aceves et al., (1986)), these become

$$G_{n+1}(\underline{x},o) = \sqrt{T}\, A(\underline{x}) + R\, e^{i\phi_0}\, G_n(\underline{x},L_1) \qquad (2)$$

This boundary condition (2), together with the nonlinear PDE (eqn. (1)), define the infinite dimensional map of interest. The scaled input pump intensity, $a(\underline{x}) = \sqrt{T} A(\underline{x})$, will typically be a gaussian shape or a plane wave $(a(\underline{x}) = a)$. The mirror intensity reflection coefficient R accounts for losses at the input and output mirrors and is the primary source of dissipation in the map ($T = 1-R$ is the intensity transmission coefficient); the other source being diffraction losses. The linear phase shift ϕ_0 allows for a mismatch of the pump wavelength from the empty resonator transmission peak.

Equation (2) serves to generate new initial data for the nonlinear PDE (eqn. (1)) on each circuit of the resonator. Our interest lies in large n asymptotic states of these maps and they are generated according to the following prescription: (1) Specify the initial data $(G_0(\underline{x},o) = \sqrt{T}A(\underline{x}))$ at the nonlinear medium input ($z = 0$) to start the first circuit $n = 1$. Solve the nonlinear PDE (eqn. (1)) to generate $G_0(\underline{x},L_1)$. (2) Use the boundary condition (eqn. (2)) to reconstitute the new initial data $(G_1(\underline{x},o)$ for the nonlinear PDE. (3) Use this new initial data to compute $G_1(\underline{x},L_1)$ and repeat (2) incrementing n each time.

We close this section with a brief overview of the plane wave approximation to the infinite dimensional map. This proves useful in orienting our study of the more complicated infinite dimensional problem and moreover allows us to highlight, in relatively simple language, some of the interesting consequences of nonlinearity in this optics problem. Assuming an infinite plane wave in the transverse dimension, the transverse laplacian term drops out of eqn. (1) and the resulting ordinary differential equation may be directly integrated. Substituting the resulting solution in the cavity boundary conditions we end up with the following map for the intracavity plane wave complex amplitude,

$$\begin{aligned} g_{n+1} &= a + Rg_n e^{i[\phi_0 + ip/2\, N(g_n g_n^*)]} \\ &= F(g_n) \end{aligned} \tag{3}$$

We use little $g(z)$ to distinguish the plane wave solution from the full transverse solution $G(\underline{x},z)$ in subsequent discussions. The Jacobian of the map is $R^2 < 1$; thus elemental areas in the complex g_n plane are contracted under its action. Also, all points are contracted into a finite region of the g_n plane and therefore we expect the large n motion to live on (i) fixed points, (ii) periodic points, or (iii) chaotic attractors. The map itself is a superposition of the three most basic operations that one can apply to a vector in the complex plane, a rotation through $\phi_0 + pN(g_n g_n^*)/2$, followed by a contraction R and finally a translation a. These actions are schematically depicted in Figure 2.

Fixed points of the map are determined in the usual manner by setting $g_n = g_{n+1} = g$ and solving the resultant transcendental equation for g. For suitable choices of parameters the map can exhibit hysteresis effects involving 3,5,7, ... fixed points. A typical hysteresis loop displaying 3 fixed points is sketched in Figure 3 for the case of a saturable nonlinearity. This figure will provide an invaluable reference point for subsequent discussions on fixed points of the infinite dimensional map. The middle branch fixed point is always unstable (a saddle point) whereas the lower and upper branches may, in principle be either stable or unstable. Ikeda (1979,1980) discovered that the stable fixed points could go unstable under larger stresses, initiating period doubling cascades to a chaotic attractor. A detailed mathematical topological analysis of this map may be found in Hammell et al. (1985). An extensive literature now exists on various aspects of this map and the reader is referred to optics literature.

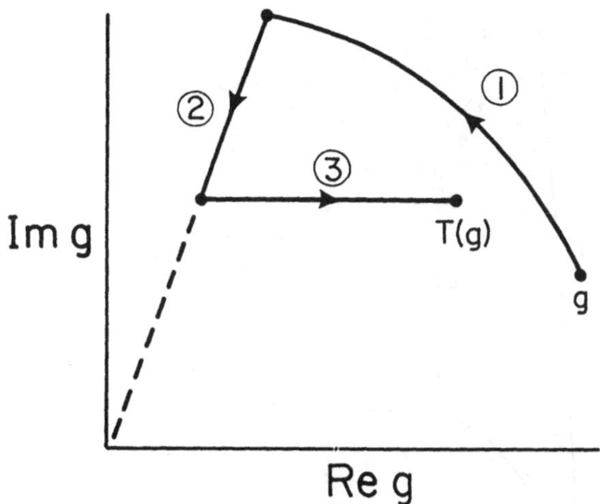

Figure 2. A sketch of the action of the plane wave optics map (eqn. (3)) in the complex plane. (1) a rotation, (2) a contraction and (3) a translation.

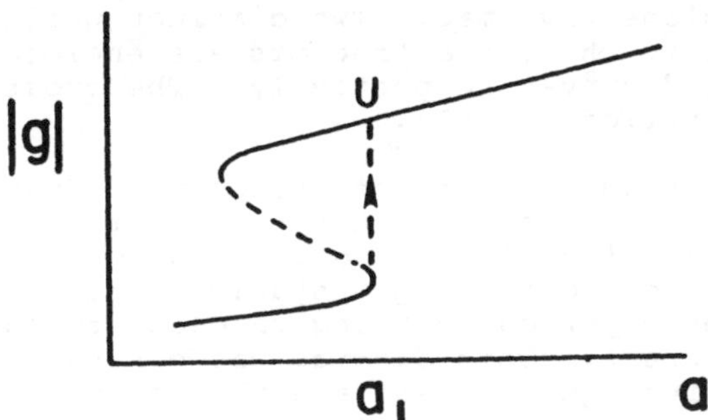

Figure 3. A hysteresis loop showing three plane wave fixed points of eqn. (3). This figure serves as a useful reference for later discussions on the infinite dimensional map.

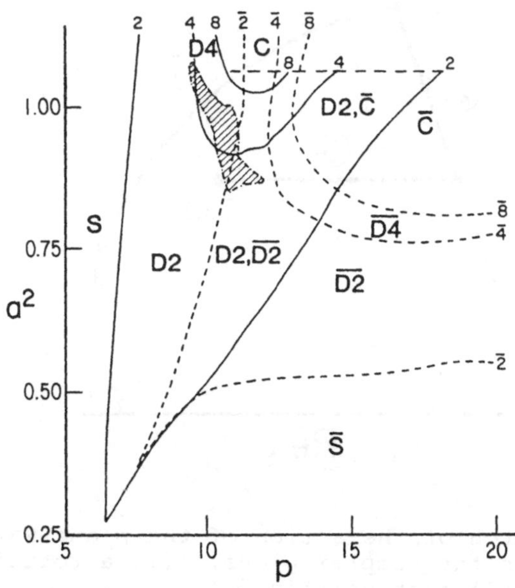

Figure 4. Numerically generated two parameter (a^2,p) bifurcation diagram indicating some of the complex dynamical behaviour of the plane wave map. Two distinct period doubling sequences to chaotic attractors are enclosed within the solid and dashed curves respectively. The cross-hatched region encloses a period 6 cycle.

Figure 4 illustrates just a part of the complex dynamical behaviour of the map on the lower hysteresis branch (Moloney (1984), Mandel et al. (1983)). This bifurcation diagram shows the existence of coexisting unstable attractors which independently undergo period doubling cascades to chaos in (a,p) - parameter space. This diagram was generated by numerically iterating eqn. (3) and a number of apparent anomalies were observed. The topological analysis by Hammel et al. (1985) has clearly isolated the underlying mechanism for these apparent anomalies and established this map as a fundamentally important mathematical object on a par with the better known Henon map. Finally, Figure 5 depicts a chaotic attractor in the region labelled C in figure 4. As we shall find in subsequent sections, very dramatic instability effects arise close to the initial instability boundary in Figure 4 at stresses much lower than those needed to generated these plane wave instabilities.

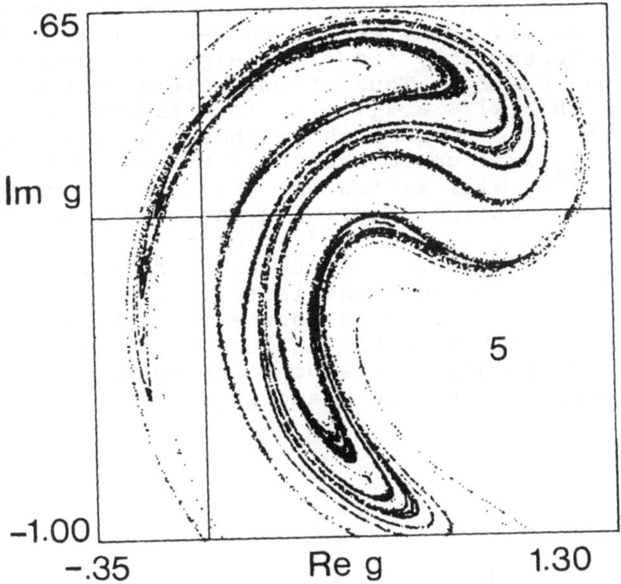

Figure 5. A chaotic attractor within the region labelled C in Figure 4.

3. FIXED POINTS OF THE MAP: NUMERICS

3.1. One transverse dimensional problem

We turn to the original problem (eqns. (1) and (2)) with one dimensional transverse effects included by considering an input field a(x) with a gaussian-like transverse profile,

$$2i \frac{\partial}{\partial z} G_n + \frac{1}{f} \frac{\partial^2}{\partial x^2} G_n + N(G_n G_n^*) G_n = 0 \tag{4}$$

$$G_n(x,0) = a(x) + Re^{i\phi_0} G_{n-1}(x,p) \quad . \tag{5}$$

Equations (4) and (5) have been rescaled to maintain consistency with McLaughlin et al., (1983). Clearly this infinite dimensional map possesses a wide variety of potential responses depending upon parameter values. We restrict these by focussing our attention (for the most part) on large Fresnel numbers ($F \simeq 100 \gg 1$) and selecting the parameters ϕ_0, p and R in regions where the plane wave map has a hysteresis diagram such as Figure 3.

In equation (4) the only coupling of a transverse segment of the beam profile to its neighbours occurs through the Laplacian $f^{-1}\partial_{xx}$. For large Fresnel number,

$f^{-1}\partial_{xx}\, a(x) \ll a(x)$ and this coupling can be initially neglected. Initially then, each transverse segment of the profile acts independently from its neighbours according to a local plane wave theory. Thus, those points on the gaussian profile for which $a(x) > a(x_+) = a_1$ (see Figure 3) will switch up to the upper branch while those parts for which $a(x) < a(x_\pm) = a_1$ will go to the lower branch. The centre of the beam profile will switch up, while its wings will not. For the saturable nonlinearity, with parameter values set at ($F = 100$, $p = 2$, $\phi_0 = .4$, $|a(0)|^2 = .0375$), this situation is shown in Figure 6. The Fresnel number F is related to f by $F = \ell n2/4\pi f$. There we show the initial gaussian profile, and the output profile after 23 passes through the nonlinear medium. Notice that the centre of the profile has switched to the upper branch while the wings have switched to the lower branch according to the plane wave theory.

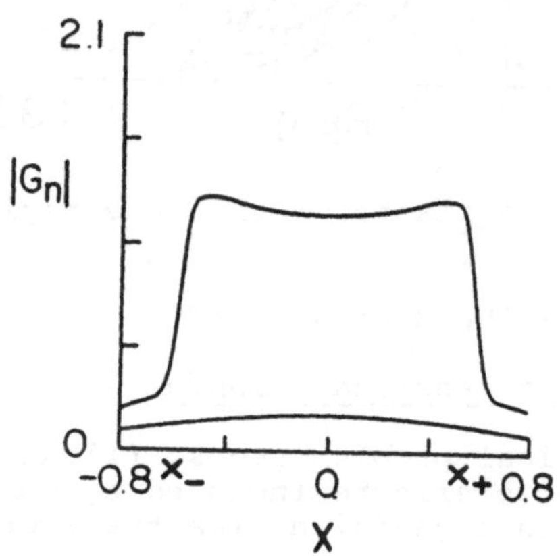

Figure 6. Development of sharp gradients from an initially broad smooth gaussian transverse profile (lower trace). The gradients develop at the spatial locations $x_\pm$ where the local amplitude $a(x) = a_1$, where a_1 is the plane wave switch-up point in Figure 3.

However, now the two outer edges of the profile possess a steep gradient and, near $x \simeq x_\pm$, $f^{-1}\, \partial_{xx}G_n$ is no longer negligible. The plane wave approximation is no longer valid. Numerical experiments (which solve the full partial differential equation) show what happens during this stage of the evolution. At the edges $x_\pm$ narrow spatial rings of width $\Delta x = O(1/\sqrt{f})$ are generated (Fig. 7). These narrow rings eventually fill out the region between x_- and x_+. Once these spatial rings form and fill up the transverse profile, they persist and describe the large n asymptotic response of the infinite dimensional map. These rings can become the steady state response of the system; that is, they can be stable fixed points of the infinite dimensional map.

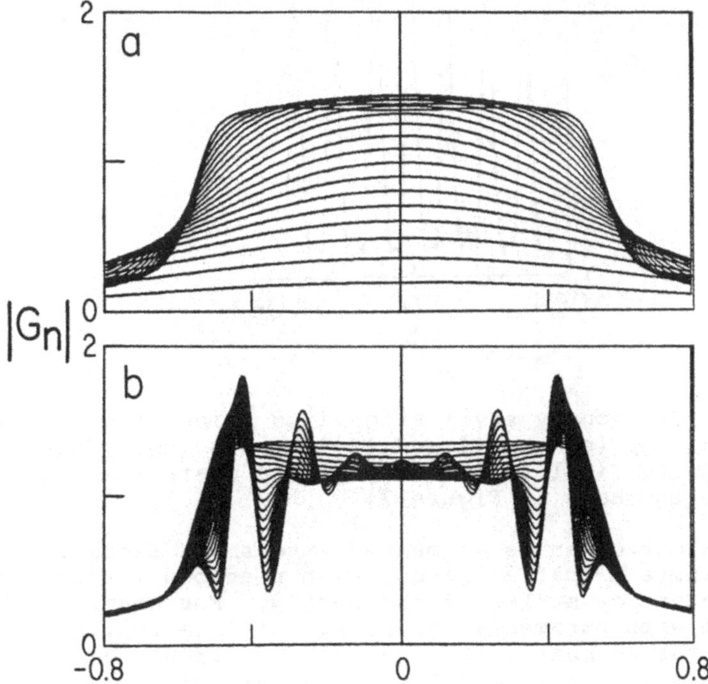

Figure 7. (a) Detailed development of the switch-up process to the stage depicted in Figure 6. The transverse profile shown at successive passes (n values) switches in the centre and approaches the quasi-plane wave fixed point at the wings. (b) Initial stages in the formation of solitary waves at the outer edges ($x = x_{\pm}$) and their progression towards beam centre.

For example, a fixed point does emerge from the transient pictured in Figure 7. This fixed point, as shown in Figure 8 at the 200th resonant pass, is a seven stationary ring pattern whose rings sit on a broad background or shelf which can be identified with the lower branch fixed point of the plane wave case.

The number of rings can be controlled. For the saturable nonlinearity, we have observed 1,3,5,7 stationary rings. The actual number of rings seems to be a function of the transient shape realised after ~ 20 passes; it is proportional to $\sqrt{f}$ and also increases with $|a(0)|^2$. Our numerical calculations show that the upper branch of the

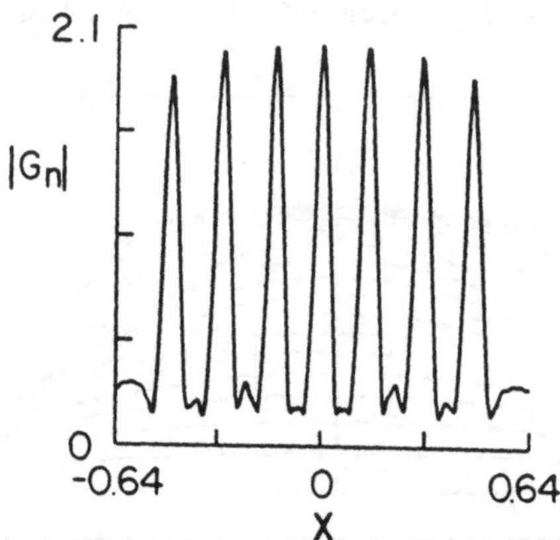

Figure 8. Stationary seven ring fixed point of the infinite dimensional map (eqns. (4) and (5)) at n = 200. This represents the final large n asymptotic state emerging from the evolution shown in Figure 7.

hysteresis curve can be segmented into small bands in $|a(0)|^2$, where n spatial rings, with n an odd integer, are the stable steady states of the system. For example, the experiment with parameter values set at (F = 100, p = 2, ϕ_0 = .4), the segment $.008 < |a(0)|^2 < .018$ has a fixed point which consists of a single "ring", whose spatial profile has an amplitude which increases and a width ($\sim 1/\sqrt{f}$) which decreases as $|a(0)|^2$ increases from .008 to .018. Increasing $|a(0)|^2$ further, for example to $|a(0)|^2 = .025$, so that it lies well beyond the switch-up point (a_1 in Figure 3) produces a stationary three ring structure.

Between the n = 1 and n = 3 stationary ring regions there exists a finite range of $|a(0)|^2$ where one observes a slow recurrent oscillation between one and two or two and three ring patterns. Even numbers of spatial rings can appear initially, but they continue to oscillate in a complicated manner on further circuits of the resonator. Whether the rings become stationary or not seems to depend on whether an odd integral number rings of width $\sim 1/\sqrt{f}$ can fit into the total area of the switched on portion of the beam (see Figure 6). The relative disposition on the n-ring stationary transverse spatial structures on the upper branch of the hysteresis is summarised in Figure 9.

We have carried out the following numerical experiment to establish the role of the external pump and dissipation in stabilising these transverse ring structures. The stationary rings which developed after 200 resonator passes were taken

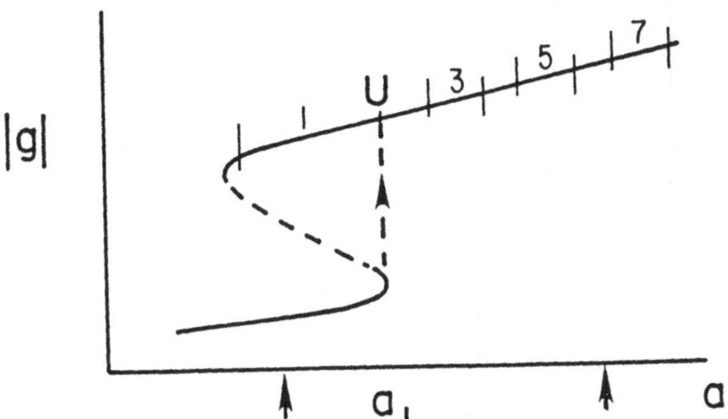

Figure 9. Location of stationary n-ring (n = 1,3,5,7) fixed points of eqns. (3) and (4) on the upper branch of the corresponding hysteresis loop. The regions in between represent slowly changing dynamical states as discussed in the text.

as initial values to the nonlinear evolution equation (4) and propagated down a long tube. After a short distance, of the order of a few medium lengths in the resonator, the rings were observed to oscillate up and down in amplitude about their original stationary values. This demonstrates immediately that the map (eqn. (4) and (5)) acts to freeze out these ring structures. For three rings at $|a(0)|^2 = .025$ [$F = 100$, $p = 2$, $\phi_0 = .4$] we observed the following behaviour. Initially the rings appear to oscillate up and down but do not attain any noticeable transverse velocities. They appear to be trapped by the broad shelf which represents the quasi-plane-wave lower branch fixed point. On further propagation the shelf becomes modulated and low amplitude rings develop. Once these are well formed, the two large amplitude outer rings in the triplet begin to slowly propagate outwards, interacting nonlinearly with and passing through their low amplitude neighbours. This behaviour is reminiscent of soliton propagation. We tracked the evolution until both outer rings reached the "1/e" value ($x = \pm 1$) of the original input gaussian beam. The centre ring just oscillates up and down. These experiments indicate that the rings are independent structures which are locked together as a consequence of the pump and dissipation terms in the map.

As mentioned earlier, the rings sit on a broad background which, since F is large, should have the height of the plane wave lower branch fixed point. To verify this, we consider the numerical experiment with parameter values ($F = 100$, $|a(0)|^2 = .008$, $p = 2$, $\phi_0 = .6$). For these values of p and ϕ_0, the plane wave hysteresis curves for both the saturable and Kerr nonlinearities are shown in Figure 10. Using these

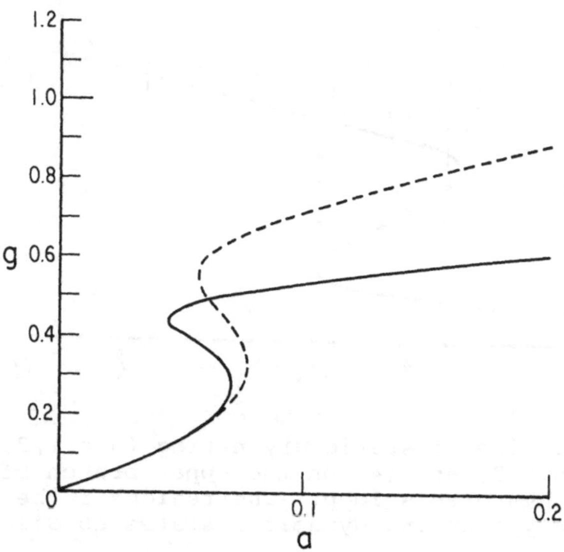

Figure 10. Plane wave hysteresis curves for a saturable (solid) and Kerr (dashed) nonlinearity

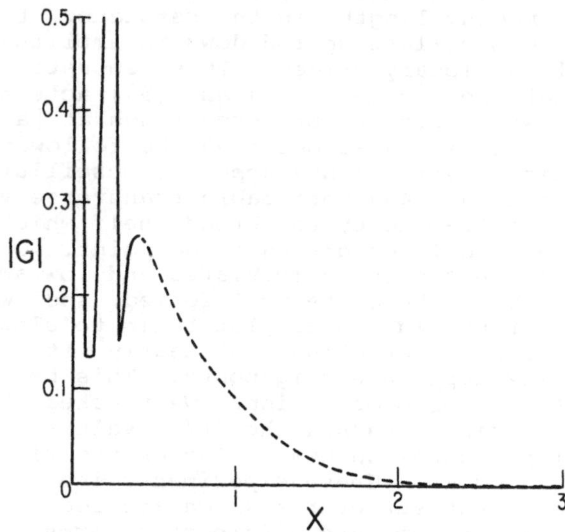

Figure 11. Comparison of the lower branch plane wave fixed points from Figure 10 with the shelf of the infinite dimensional map three ring fixed point. This curve represents a blow-up of part of the three ring fixed point computed for a saturable nonlinearity and shown in the next figure.

hysteresis curves, we compute how the wings of the transverse profile should behave. The prediction is compared with the actual transverse experiments in Figure 11 and shows perfect agreement.

The switch up mechanism is primarily controlled by the size of the Fresnel number F. This latter quantity also plays a major role in determining the width of the upper branch rings. When F is large, these rings are skinny and tall. For example, with $F = 100$, $|a(0)|^2 = .025$, $p = 2$, $\phi_0 = .4$, the situation is shown in Figure 12 for the three ring fixed point.

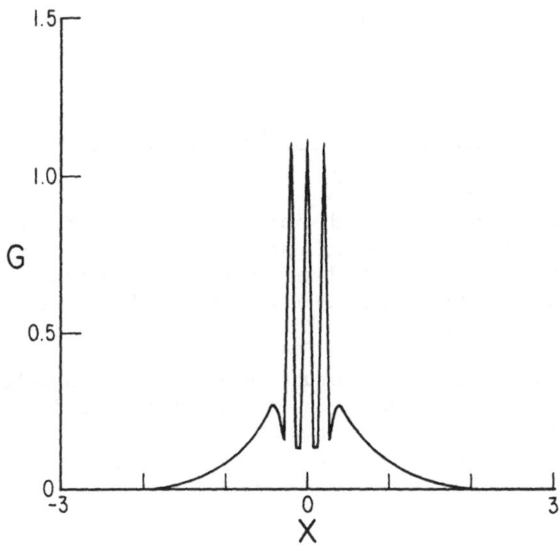

Figure 12. Three ring fixed point of the infinite dimensional map (eqns. (4) and (5)) for a saturable nonlinearity. This fixed point lies in the parameter region labelled by 3 in Figure 9. [$F = 100$, $a(0)^2 = .025$, $p = 2$, $\phi_0 = .4$].

We close this section by describing some of the differences that we have observed when the saturable is replaced by the Kerr nonlinearity. Generally, both the evolution of the beam and the final asymptotic states are much more sensitive to the pumping amplitude in the Kerr case. First, it is difficult to achieve a single ring asymptotic state in the Kerr case. The only region on the upper branch where a single ring shape appears to arise is at the extreme left near the switch down point. In Figure 13 we show one such single ring profile for [$F = .8$, $p = 1.5$ $\phi_0 = .4$, $|a(0)|^2 = .003$]. Secondly, fixed points are more difficult to achieve. As $|a(0)|^2$ is increased, extra rings do appear, but the beam approaches an oscillatory state rather than a fixed point. This state resembles an exact multisoliton

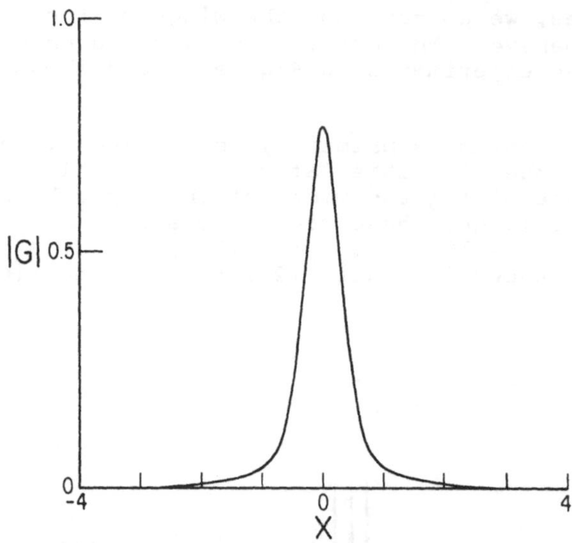

Figure 13. Example of a fixed point of the infinite dimensional map for a Kerr nonlinearity. This fixed point occurs at the extreme left end of the corresponding hysteresis close to the switch-down point [$F = .8$, $p = 1.5$, $\phi_0 = .4$, $|a(0)|^2 = .003$].

solution of the integrable (Kerr) nonlinear Schrodinger equation. It should be constrasted with the multi-ring fixed points which arise in the saturable case and which appear to be phase locked individual entities. To emphasise the distinction we propagated both asymptotic states down an extended tube. In the Kerr case, the asymptotic state of the resonator persisted as a coherent transverse structure which did resemble (in one case at least) an analytically generated 2-soliton wave form. On the other hand, in the saturable case, the individual rings did not remain locked together as they propagated down the long tube; instead, they drifted apart at different velocities.

We can conclude from these numerical studies that the asymptotic dynamic states for the field on the upper hysteresis branch differ significnatly for saturable and Kerr nonlinearities. In the Kerr case, fixed points are rare; the asymptotic states tend to be oscillatory. They also depend rather sensitively on the pumping amplitude. The saturable case has more controlled, more stable asymptotic responses.

3.2 Numerical comparison of transverse profiles with solitary waves

In this subsection we fit solitary waves to the transverse profiles which are fixed points of the infinite dimensional map. These fixed point profiles are generated by solving eqns. (4) and (5) numerically. First, we describe the saturable case. Typical results are depicted in Figure 14a. There we show a transverse profile generated by solving eqns. (4) and (5) with parameters set at $F = 100$, $p = 2$, $\phi_0 = .4$, $|a(0)|^2 = .008$ (for single soliton). The profile shown is that of the 200th pass, by which time all transients have died out and the profile is certainly a fixed point. In this case, the profile consists of a single pulse on top of a flat background. The pulse is accurately represented by a solitary wave as the figure shows. To obtain that fit, we select the amplitude parameter λ to agree with the numerical peak and solve equation (7), see section 4, with appropriate boundary conditions. The wings of the actual profile do not approach zero as does the solitary wave; rather, as established in section 3 they approach the lower branch height of the local plane wave hysteresis curve.

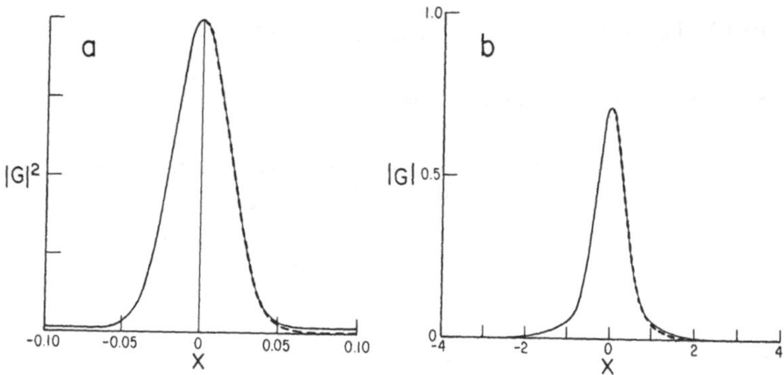

Figure 14. Comparison of numerically generated fixed points of the infinite dimensional map with solitary wave shapes. (a) Saturable nonlinearity. [$F = 100$, $p = 2$, $\phi_0 = .4$, $|a(0)|^2 = .008$]. (b) Kerr nonlinearity. [$F = 0.8$, $p = 1.5$, $\phi_0 = .4$, $|a(0)|^2 = .003$]. The dashed curves represent the solitary shapes in both cases.

Turning to the case of Kerr nonlinearity, it is more difficult to generate fixed points. One such fixed point is shown in Figure 14b where we choose $F = 0.8$, $|a(0)|^2 = 0.003$, $p = 1.5$, $\phi_0 = .4$. The dashed curve is the exact soliton solution which has the same amplitude as the observed fixed point. The two curves agree very well except at the wings. As mentioned above, if the input intensity (stress) is raised, the wings begin to grow and interact with the central peak. Then the whole profile starts to oscillate. With these parameter values we have not observed multi-ring fixed points like those obtained in the saturable case; rather, the oscillation seems to persist indefinitely.

The numerical experiments just described show that the central part of the transverse fixed points are well approximated by solitary waves. However, the correct amplitude parameter is chosen by a numerical fitting procedure. Next, in section 4, we determine the correct amplitude parameter analytically.

4. FIXED POINTS OF THE MAP: ANALYSIS

4.1 Solitary waves

Known theory for the nonlinear wave equation (4) indicates that the transverse rings should be solitary waves. A particular solitary wave is a solution of (4) in the form ($y = \sqrt{f}\, x$).

$$G_s(y,z;\lambda) = S(\lambda y;\lambda)e^{i[(\lambda^2-1)\frac{z}{2}]} \quad , \tag{6}$$

where $S(\theta;\lambda)$ is a real, even solution of

$$S_{\theta\theta} - S + \frac{1}{\lambda^2}[1 + N(S^2)]S = 0 \tag{7}$$

which vanishes as $\theta = \lambda y \to \infty$. The general solitary wave is a four parameter family of solutions of (4),

$$G_s(y,z;\lambda,\gamma,a,v) = S[\lambda(y-a-vz);\lambda]\, e^{i[vy+(\lambda^2-v^2)\frac{z}{2} + \gamma]} \tag{8}$$

which can be obtained from the particular solution (6) by using the symmetries of phase, translation, and Galilean invariance. In this work it will be sufficient to consider the two parameter family

$$G_s(y,z;\lambda,\gamma) = S(\lambda y;\lambda)e^{i(\lambda^2\frac{z}{2} + \gamma)} \quad . \tag{9}$$

because the transverse profiles are symmetric about the beam axis. The parameter λ determines the amplitude and width of the solitary wave, while γ determines its phase. In the case of the Kerr nonlinearity $[1 + N(S^2) = 2S^2]$, the solitary wave takes the explicit form

$$S(\theta;\lambda) = \lambda \operatorname{sech} \theta \quad . \tag{10}$$

Here λ certainly determines the amplitude and width of the solitary wave. For more general nonlinearities, one must study the differential equation (7). We illustrate for the saturable case where

$$1 + N(S^2) = 1 - \frac{1}{1+2S^2} = \frac{2S^2}{1+2S^2} . \tag{11}$$

Introducing a potential V(S), which for this saturable case takes the form

$$V(S) = \left(\frac{1}{\lambda^2} - 1\right) \frac{S^2}{2} - \frac{1}{4\lambda^2} \ln(1 + 2S^2) , \tag{12}$$

the differential equation (7) may be rewritten as

$$S'' = - \frac{\partial}{\partial S} V(S) . \tag{13}$$

This equation has the immediate "energy integral"

$$\frac{1}{2} S'^2 = E - V(S) . \tag{14}$$

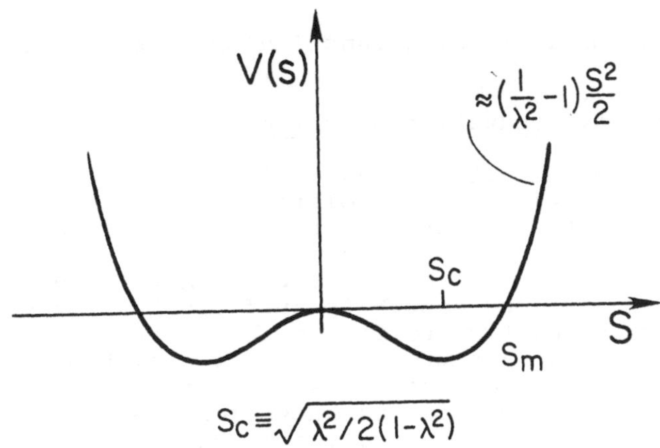

Figure 15. Sketch of the potential V(S) for the saturable nonlinearity. The solitary wave amplitude S_m is determined from the condition $V(S_m) = 0$ as indicated in the sketch.

The potential V(S) is sketched in Figure 15. From this sketch of V, we see that E must be chosen as zero if we are to satisfy the boundary condition at $\theta = \infty$, and that the parameter λ may take any value in the range $0 < \lambda \leqslant 1$. (No solitary wave exists for $\lambda > 1$). With these considerations the qualitative shape of the solitary wave $S(Y;\lambda)$ is sketched

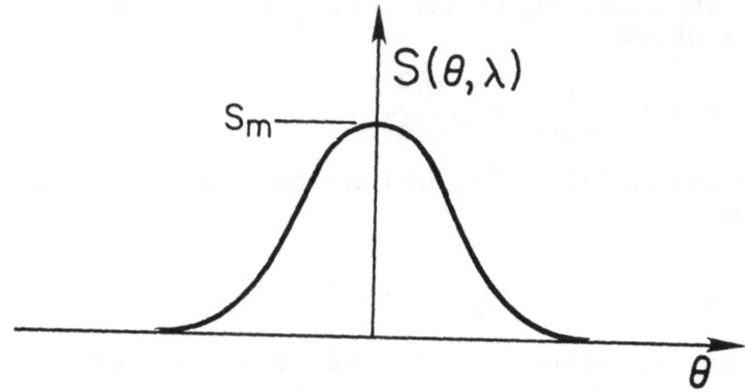

Figure 16. The solitary wave shape derived from the potential V(S) in Figure 15.

in Figure 16. The amplitude $S_m = S_m(\lambda)$ is determined from

$$(1-\lambda^2)\ S_m^2 = \frac{1}{2}\ \ell n(1 + 2S_m^2) \quad , \tag{15}$$

and is a monotone increasing function of λ as λ runs from 0 to 1.

4.2 Solitary wave reduction of the map

We return to the infinite dimensional map (4) and (5) and consider the possibility of a solitary wave reduction. Given the laser field at the entry point z = 0 to the nonlinear medium, we would like to predict which solitary wave emerges at the exit point z = p. For this problem to have an answer, the medium must be long enough (p sufficiently large) that the nonlinearity has time to filter the laser field into its asymptotic solitary wave profiles. We restrict our attention to such sufficiently long cavities. Under this restriction a global answer to our problem is known for the Kerr nonlinearity. That is, given any initial data, one can predict, using the inverse scattering transform, exactly which solitons emerge at the end of a Kerr medium. Unfortunately, this mathematical transform method does not apply to more realistic saturable media, and we must content ourselves with a more local problem. For initial data which is close to a given solitary wave, can we predict which modified solitary wave emerges? That is, what are the values of its parameters?

In general, they are not equal to the input parameters because some of the solitary wave which emerges is hidden in the perturbation a(x) of the initial data. However, as this perturbation is assumed to be small (of order ε), we can assume that the parameters of the solitary wave which emerges from the end of the cell are close to those of the solitary wave at the cell entrance. Formally, then, we linearise about the solitary wave which eventually emerges on each pass and the problem is to determine its parameters λ and γ which are, as yet, unknown. Mathematical details of the solitary wave projection formalism are given by Adachihara et al. (1986). Here we simply provide the end result of the derivation.

Motivated by the discussion of the preceding paragraph we rewrite the infinite dimensional map (eqns. (4) and (5)) in the following form

$$2i \frac{\partial}{\partial z} G_{n+1} + G_{n+1,yy} + N(|G_{n+1}|^2)G_{n+1} = 0 \tag{16}$$

$$G_{n+1}(y,0) = a(y) + Re^{i\phi_0} e^{i[\frac{\lambda_n^2-1}{2} p + \gamma_n]} S(\lambda_n y;\lambda_n) \tag{17}$$

In writing (16,17), we have assumed that the output on the nth pass down the nonlinear medium is a pure solitary wave. All other modes, such as radiation, have been neglected. In addition, we have elected to examine an output which is symmetric about the y = 0 axis. Thus, the parameters a and v may be ignored by symmetry considerations. Since $R \approx 1$ and $a(y) \ll 1$, initial data (17) may be treated as a small perturbation of a solitary wave. (Here the phase factor $(\phi_0 + (\lambda^2{}_n - 1)\, p/2 + \gamma_n)$ merely changes the phase in the solitary wave). We now use the general formalism developed by Adachihara et al. to predict values for the parameters of the solitary wave which emerges after the next pass. The end result is the following map:

$$\begin{aligned} \langle S_{n+1}, S_{n+1}\rangle &= \langle A_{n+1}, S_{n+1}\rangle \cos \gamma_{n+1} \\ &\quad + R \cos\Gamma_{n,n+1} \langle S_{n+1}, S_{n,n+1}\rangle \\ 0 &= -\langle A_{n+1}, P_{n+1}\rangle \sin\gamma_{n+1} \\ &\quad + R \sin \Gamma_{n,n+1} \langle \rho_{n+1}, S_{n,n+1}\rangle \end{aligned} \tag{18}$$

where

$$A_{n+1} = a(\theta/\lambda_{n+1})$$

$$S_{n+1} = S(\theta;\lambda_{n+1})$$

$$\Gamma_{n,n+1} = \phi_0 + (\gamma_n - \gamma_{n+1}) + \frac{p}{2}(\lambda_n^2 - 1) \tag{19}$$

$$S_{n,n+1} = S\left(\frac{\lambda_n \theta}{\lambda_{n+1}}\right); \lambda_n$$

$$\rho_{n+1} = \frac{1}{\lambda_{n+1}} \theta (S_\theta(\theta; \lambda_{n+1}) + \frac{\partial}{\partial \lambda} S(\theta, \lambda) \Bigg|_{\lambda = \lambda_{n+1}}$$

Equation (18) defines a two dimensional real map on the solitary wave parameters (λ, γ):

$$(\lambda_n, \gamma_n) \rightarrow (\lambda_{n+1}, \gamma_{n+1}) \quad . \tag{20}$$

We have used solitary wave perturbation theory to reduce the infinite dimensional map (4,5) to a two dimensional one (18). Note that this map is implicit with the unknown solitary wave amplitude (λ_{n+1}) and phase (γ_{n+1}) appearing on both sides of the equation. Moreover, the expressions in angle brackets involve projections of solitary waves which depend on these unknown parameters. For example,

$$\langle A_{n+1}, S_{n+1} \rangle = \int_{-\infty}^{\infty} a(\theta/\lambda_{n+1}) \, S(\theta; \lambda_{n+1}) \, d\theta \quad . \tag{21}$$

The equation for the fixed points (λ, γ) is much simpler than the map itself,

$$\begin{aligned} (S,S) &= (A,S) \cos \gamma + R \cos \Gamma \, (S,S) \\ 0 &= -(A,\rho) \sin \gamma + R \sin \Gamma \, (\rho, S) \quad , \end{aligned} \tag{22}$$

where

$$S = S(\theta; \lambda)$$

$$A = a(\theta/\lambda)$$

$$\Gamma = \phi_0 + \frac{p}{2} (\lambda^2 - 1)$$

$$\rho = \frac{1}{\lambda} \theta \, S_\theta + \frac{\partial}{\partial \lambda} S(\theta; \lambda) \quad .$$

The main result of this section is the map

$$(\lambda_n, \gamma_n) \rightarrow (\lambda_{n+1}, \gamma_{n+1})$$

on the amplitude and phase parameters which is given by (18). The fixed points of this map satisfy (22); thus (22) predicts the parameter values for the solitary wave which finally emerges after many passes through the nonlinear medium.

In general equation (22) for the fixed points is quite implicit. First, we solved it numerically for the saturable case. Typical results ar shown in Figure 17. We emphasise that there are no free parameters in this theory. The theory rigorously predicts the amplitude of the solitary wave which emerges. The results (Figure 17) are very accurate.

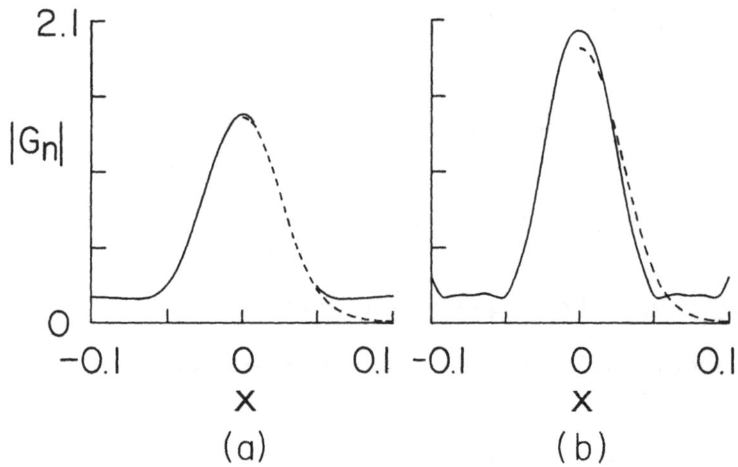

Figure 17. Comparison of the solitary wave fixed points (dashed curves) of the reduced map (eqn. (22) with the numerically computed fixed points of the infinite dimensional map (eqns. (4) and (5)) for the case of a saturable nonlinearity. (a) Single solitary wave on the upper part of the hysteresis loop (region 1 in Figure 29). (b) Central solitary wave of the seven-solitary wave fixed point shown in Figure 8. Perturbations due to neighbouring solitary waves in this latter picture lead to poorer agreement.

4.3 Reduced map - Kerr case

In the Kerr case, much more can be done analytically, primarily because of the explicit formula for the solitary wave

$$S(\theta;\lambda) = \lambda \operatorname{sech} \theta \quad .$$

Using this formula we place the reduced map (18) in the form

$$\lambda_{n+1} = A_s(\lambda_{n+1})\cos \gamma_{n+1} + R\, B_s(\lambda_n/\lambda_{n+1}) \cos(\Gamma_{n,n+1})\, \lambda_n$$

$$0 = -A_\rho(\lambda_{n+1}) \sin \gamma_{n+1} + RB_\rho(\lambda_n/\lambda_{n+1}) \sin(\Gamma_{n,n+1})\, \lambda_n \tag{23}$$

where

$$A_s(\lambda_{n+1}) = \frac{1}{2} \int \operatorname{sech} \theta \; a\left(\frac{\theta}{\lambda_{n+1}}\right) d\theta$$

$$A_\rho(\lambda_n) = \int (\theta \operatorname{sech} \theta)_\theta \; a\left(\frac{\theta}{\lambda_{n+1}}\right) d\theta$$

$$B_s(\lambda_n/\lambda_{n+1}) = \int \operatorname{sech} \theta \operatorname{sech} \left(\frac{\lambda_n}{\lambda_{n+1}} \theta\right) d\theta$$

$$B_\rho(\lambda_n/\lambda_{n+1}) = \int (\theta \operatorname{sech} \theta)_\theta \operatorname{sech} \left(\frac{\lambda_n \theta}{\lambda_{n+1}}\right) d\theta$$

$$\Gamma_{n,n+1} = \phi_0 + (\gamma_n - \gamma_{n+1}) + \frac{p}{2}(\lambda_n^2 - 1) \quad .$$

Map (23) should be compared with the plane wave map for the Kerr nonlinearity

$$g_{n+1} = a + Re^{i[\phi_0 + \frac{p}{2}(2|g_n|^2-1)]}\, g_n \quad , \tag{24}$$

which may be rewritten in the form ($g = \lambda e^{i\gamma}$)

$$\lambda_{n+1} = a \cos \gamma_{n+1} + R \cos(\Gamma_{n,n+1})\, \lambda_n$$

$$0 = -a \sin \gamma_{n+1} + R \sin(\Gamma_{n,n+1})\, \lambda_n \quad . \tag{25}$$

In this Kerr case, the map on solitary wave parameters, (eqn. (23)) and the plane wave map are very similar. The main difference is that constants in the plane wave case are replaced by projections over solitary wave profiles. These projections make the map (23) implicit in contrast to its plane wave counterpart, because of the dependence on λ_{n+1} on the right hand side. More importantly, a symmetry in the plane wave map $[a \cos \gamma_{n+1}, -a \sin \gamma_{n+1}]$ is drastically broken by these projections. To see this, realise that the solitary waves which evolve may be typically narrow when compared to the input gaussian. In this case, the projections of the gaussian a(y) can be estimated:

$$A_s(\lambda) = \frac{1}{2} \int \text{sech}^{\,1}\, a\left(\frac{\theta}{\lambda}\right) d\theta \simeq \frac{\pi}{2}\, a(0) \tag{26}$$

$$A_\rho(\lambda) = \int (\theta\ \text{sech}\ \theta)_\theta\ a\left(\frac{\theta}{\lambda}\right) d\theta \simeq 0$$

We use this calculation to introduce a third map, which we call the "constant-Kerr" case (a = a(0)):

$$\begin{aligned} \lambda_{n+1} &= \frac{\pi}{2}\, a \cos \gamma_{n+1} + R\, B_s(\lambda_n/\lambda_{n+1}) \cos(\Gamma_{n,n+1})\, \lambda_n \\ 0 &= \qquad\qquad + R\, B_\rho(\lambda_n/\lambda_{n+1}) \sin(\Gamma_{n,n+1})\, \lambda_n \ . \end{aligned} \tag{27}$$

This last map is rather easy to analyse. Its fixed points (λ, γ) satisfy

$$\begin{aligned} \lambda &= \frac{\pi}{2}\, a \cos \gamma + R(\cos\Gamma)\, \lambda \\ 0 &= (\sin \Gamma)\lambda, \quad \Gamma = \phi_0 + \frac{p}{2}(\lambda^2-1) \ , \end{aligned} \tag{28}$$

which can be solved explicitly to yield

$$\lambda = 0$$

$$\Gamma = j\pi \qquad \lambda_j = \sqrt{1 + (j\pi - RL)\frac{2}{p}} \tag{29}$$

$$\cos \gamma_j = \frac{2}{\pi a}\left[1-(-1)^j R\right] \lambda_j$$

These are sketched in Figure 18. Notice that the only a^2 dependence is a lower cut-off which guarantees $\cos \gamma_j \quad 1$. These cut-offs are the maximum-minimum responses of the plane wave map, equation (24). Notice also that each curve, except $\lambda = 0$, stands for two fixed points $(\lambda_j', \pm \gamma_j)$. As approximation (26) is removed, this degeneracy is broken and the curves develop a dependence on the amplitude a^2. These are pictured in Figure 19.

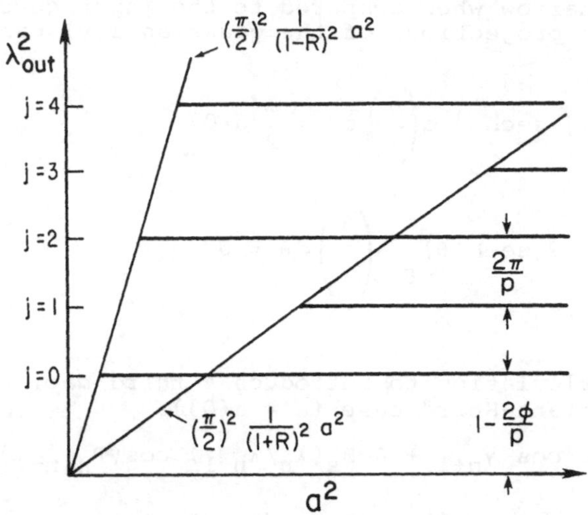

Figure 18. Fixed points of the "constant-Kerr" map (eqn. (27)). All curves except $\lambda = 0$ are doubly degenerate. Only the even j curves are physically relevant and the two lines denoting the extrema of these curves represent the maximum-minimum responses of the plane wave map.

Next, we linearise this map about a fixed point (λ_j, λ_y):

$$\begin{bmatrix} \bar{\lambda}_{n+1} \\ \bar{\gamma}_{n+1} \end{bmatrix} = T(\lambda, \gamma) \begin{bmatrix} \bar{\lambda}_n \\ \bar{\gamma}_n \end{bmatrix} \tag{30}$$

where

$$\lambda = \lambda_j \neq 0$$

$$\gamma = \gamma_j$$

$$T(\lambda, \tau) = \begin{bmatrix} \dfrac{-p\,\dfrac{\pi a}{2}\,\lambda \sin\gamma + (-1)^j \dfrac{R}{2}}{(1 + (-1)^{j+1} R/2)} & \dfrac{-\dfrac{\pi a}{2} \sin\gamma}{(1 + (-1)^{j+1} R/2)} \\ p\lambda & 1 \end{bmatrix} \tag{31}$$

Now the Jacobian at this fixed point is given by

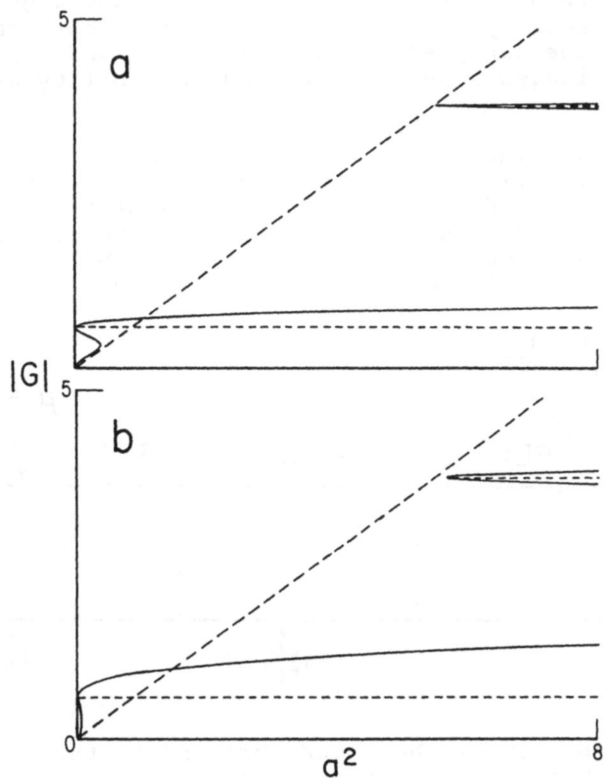

Figure 19. Fixed points of the reduced Kerr map (eqn. (23)). The degeneracy of the curves in Figure 18 is now lifted due to the finite contribution of the second projection in eqn. (26).

$$\det T(\lambda,\gamma) = \frac{(-1)^j R/2}{1-(-1)^j R/2} \tag{32}$$

This is approximately equal to R^2 if j is even and far from R^2 if j is odd. For this reason, we restrict our attention to the even values of j. For even j, the eigenvalues of T satisfy,

$$\mu^2 - \frac{1 - \frac{p\pi a}{2}\lambda \sin\gamma}{1-R/L}\,\mu + \frac{R/2}{1-R/2} = 0 \ . \tag{33}$$

As in the case of the plane wave map, these eigenvalues are either both real or are conjugates of each other. In the latter case, they always satisfy $|\mu| < 1$, and, hence, the corresponding fixed points are stable. When the eigenvalues are real, they satisfy $\mu_1\mu_2 = R/(2-R)$. They both can be less than 1 or one can pass through either +1 and -1. The latter case indicates a period doubling instability of the fixed point.

Consider a fixed point (λ_j, γ_j) and its sister $(\lambda_j; -\gamma_j)$. We call $(\lambda_j, \gamma_j > 0)$ the "lower jth branch" and $(\gamma_j; -\gamma_j)$ the "upper jth branch" because of their locations once the degeneracy is split by removing (26). One can show that the lower branch is always unstable. The upper branch has the stability depicted in Figure 20.

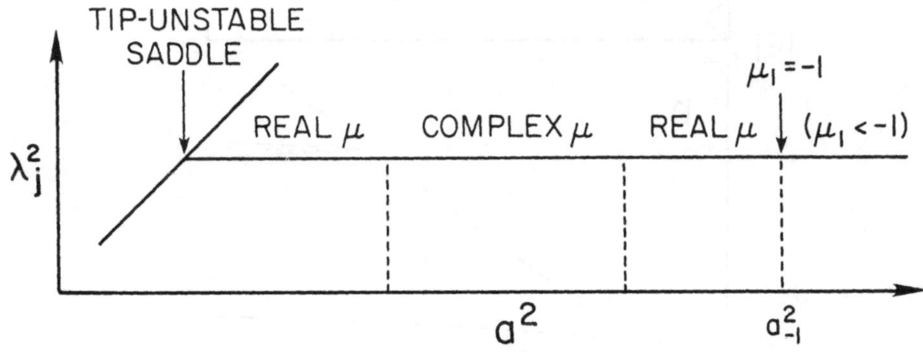

Figure 20. Stability of the "upper branch" of the degenerate set in Figure 18 as a function of increasing pump intensity a^2. The point labelled a^2_{-1} denotes the period doubling bifurcation point $\mu_1 = -1$).

A curve can be drawn, to the immediate right of which a period 2 bifurcation occurs. Some parameter values at which the period doubling bifurcation occurs are listed in Table 1.

$\phi = .4$	$p = 2\pi$	$R = .9$	$\phi = 4$	$p = 2$	$R = .9$
j=0	$a_{-1}^2 = .0506$		j=0	$a_{-1}^2 = .677$	
j=2	$a_{-1}^2 = .0260$		j=2	$a_{-1}^2 = .0867$	
j = 4	$a_{-1}^2 = 0.282$		j=4	$a_{-1}^2 = .0841$	
j = 6	$a_{-1}^2 = .0338$		j=6	$a_{-1}^2 = .0994$	
j=8	$a_{-1}^2 = .0406$				

When a is nonconstant the approximation (26) no more applies and the fixed point of the map (23) has to be evaluated numerically. To be consistent with (4), (5) we choose

$a(y) = a(0)\, e^{-\frac{\ell n 2}{4\pi F p} y^2}$. A result of such a computation is shown in Figure 21 where the solitary wave fixed point is plotted as a dashed line and, for comparison, the plane wave fixed point is also plotted as a solid line. The parameter values are F = 0.8, p = 1.5, ϕ_0 = 0.4.

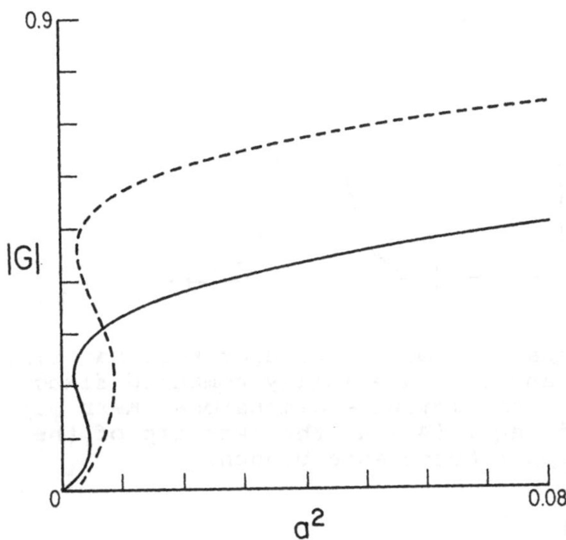

Figure 21. Comparison of the solitary wave fixed points (solid curves) of the reduced Kerr map and the plane wave fixed points (dashed curve).

Keeping the same parameter values we solved the equation numerically and compared resulting fixed points with analytical ones. One such attempt is shown in Figure 22 where the dashed curve represents predicted solitary wave fixed points. The relative error in amplitude is about 2.9%.

The discrepancy comes from two sources. One is that the observed fixed points are not pure solitons but rather a combination of a single soliton and a lower branch fixed point or wings, where the latter presumably "pushes up" the former resulting in a larger amplitude than the prediction. The other is more subtle. In deriving the map we assumed the form (17) whereby each time the boundary condition is updated

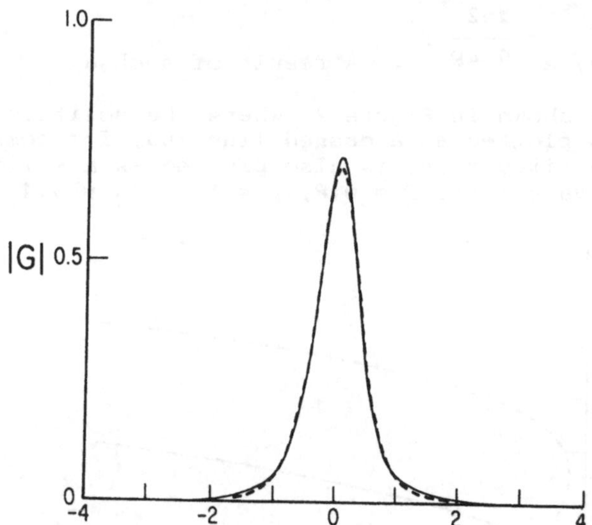

Figure 22. Comparison of the reduced Kerr map fixed point (dashed curve) and the numerically computed fixed point (solid curve) of the infinite dimensional Kerr map ($N(I) = -1+2I$ in eqn. (4)) at the left tip of the corresponding upper hysteresis branch.

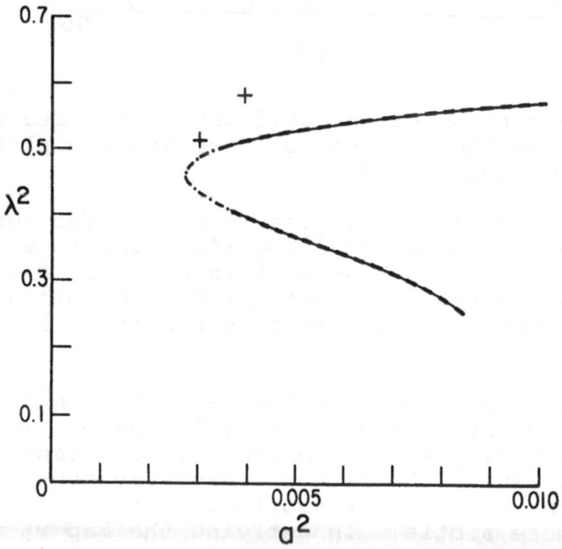

Figure 23. The reduced Kerr map fixed points (solid curve) showing the expected region of validity of the approximation leading to eqn. (23) (cross hatched region near the left tip). The crosses denote the fixed points computed from the infinite dimensional Kerr map.

by a soliton plus a small perturbation. Since the explicit expression of this perturbation term is known in our formalism, we can compute its size relative to the soliton. The computation shows that for certain parameter values its size becomes large. The region where the perturbation stays small turns out to be confined to the left most part of the hysteresis, which is shown shaded in Figure 23. The crosses in the same figure represent the observed fixed points. At the "tip" of the curve the relative size of the perturbation becomes minimum, where we expect the best fit.

Numerical experiments such as these establish that, when solitary wave fixed points occur, their amplitudes (and widths) are accurately predicted by the reduced maps, (22) in the saturable case and (23) in the Kerr case. However, the experiments also show that the stability of these fixed points is not accurately captured by these reduced maps. Accurate stability calculations require the inclusion of more degrees of freedom than just the (2 parameter) solitary wave ansatz. Stability calculations are discussed in Section 6.

5. ONSET OF MODULATION INSTABILITIES: NUMERICS

The previous section established the close connection between fixed points of the infinite dimensional map associated with the upper branch of the hysteresis loop (Fig. 3) and fixed points of a reduced map in the solitary wave parameters. Numerical results from section 3 show that there also exist on the upper branch finite bands in parameter space where the solution of the infinite dimensional map never reaches a steady state but instead undergoes slow recurrent oscillations in time. We now increase the stress parameters (specifically a(y) and p) and study the onset of instabilities on both upper and lower hysteresis branches. The oscillations associated with this latter instability have characteristic periods of the order of the cavity round trip time. The initial bifurcation, as in the plane wave map, is a period doubling one corresponding to an eigenvalue of the linearisation of the infinite dimensional map crossing through the unit circle along the negative real axis. We remark here that the infinite dimensional map allows a much broader class of bifurcation as discussed by Moloney (1984c,1986b) and Moloney et al. (1982). Throughout the present article we confine our attention to parameter choices where one might anticipate quasi-plane wave behaviour by choosing a reasonably large Fresnel number and indeed our analysis in the next section shows that the phenomena to be discussed here can be accounted for by considering transverse perturbations about a plane wave.

As in the earlier sections, we hold the parameters R and ϕ_0 fixed at .9 and .4, respectively, increase p to 6 and treat the input beam pump amplitude a(y) as our bifurcation

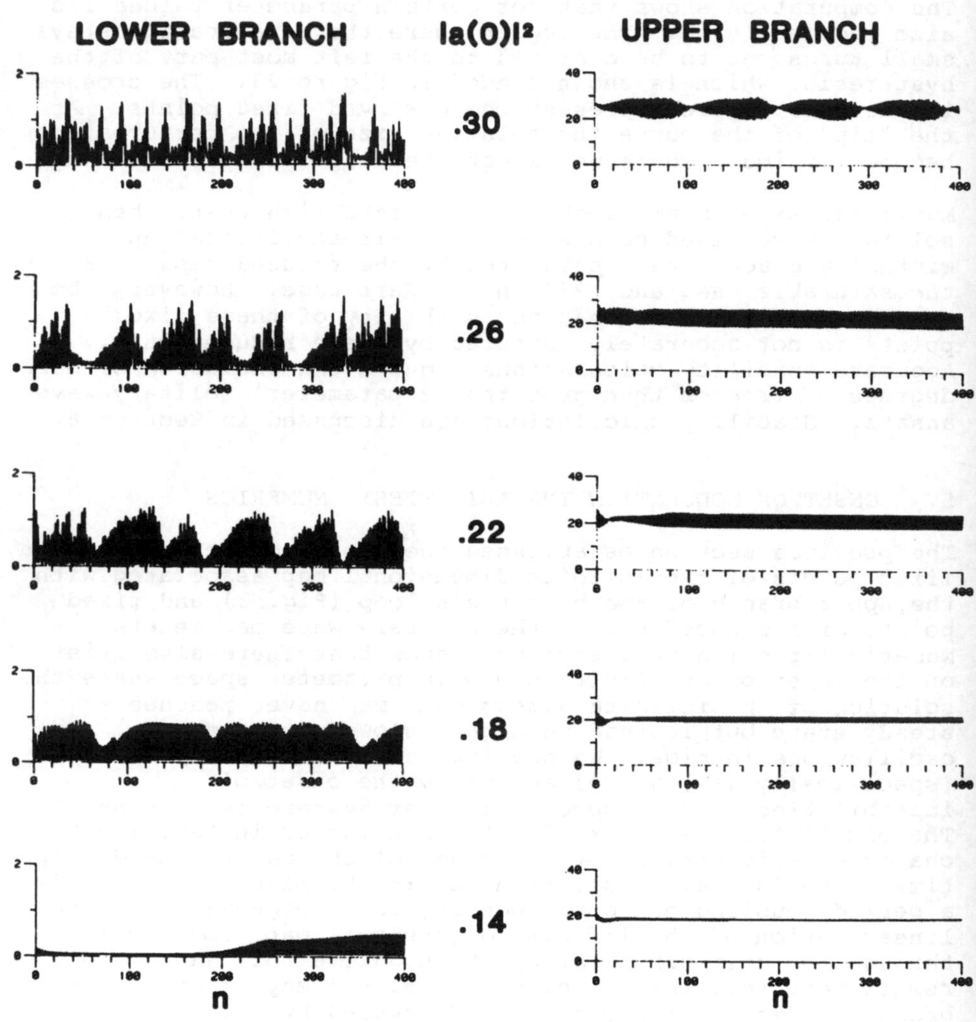

Figure 24. Time records of the centre peak amplitude $G_n(0,0)$ on both lower and upper hysteresis branches as a function of increasing pump intensity $a(0)^2$. [F = 33.3, p = 6, ϕ = .4, R = .9]. These parameters are used in all subsequent figures.

parameter. The Fresnel number F = 33.3 to ensure weak diffraction coupling. Referring to the plane wave bifurcation diagram in Figure 4 we see that we are just below the initial period doubling boundary for this latter map and are in a region of hysteresis. [Recall that the upper hysteresis branch is stable for the plane wave map [Moloney (1984), Hammel et al. (1985)]]. From the results of the earlier sections we would anticipate that the actual hysteresis curve for the infinite dimensional map would consist of a single solitary wave sitting on a broad low amplitude shelf; this low amplitude shelf itself being the lower branch quasi-plane wave fixed point. The behaviour of the solution on both branches will now be studied numerically over a range of input pump beam peak intensities $|a(0)|^2$, $(.12 < |a(0)|^2 < .32)$. This range of pump intensities encompasses the entire hysteresis region (see Figure 3). Figure 24 shows the peak amplitude, $|G_n(0,0)|$, as a function of increasing iterates of the map n over the pump intensity range $.14 < |a(0)|^2 < .30$. Time series are shown for both upper and lower branch solutions. At low intensity $(|a(0)|^2 < .12)$ on the lower branch, the solution is a stable fixed point of the map. At $|a(0)|^2 = .14$ the fixed point has lost its stability and we observe an exponential growth to a saturated period 2 oscillation for $n > 250$. The final asymptotic dynamical state of the map is therefore a period 2 oscillation. Increasing the external stress further leads to a modulation on the period 2 wavetrain at $|a(0)|^2 = .18$. This modulation manifests itself as an apparently random collapse of the saturated periodic wavetrain, approaching an amplitude close to the original fixed point, followed by an exponential growth back to the fully saturated period 2 cycle. This motion is clearly chaotic and has no analogy for a simple plane wave; indeed as Figure 4 shows the upper and lower branches of the corresponding plane wave map are stable over the entire hysteresis region. At larger stresses $(.22 < |a(0)|^2 < .30)$ the modulation collapse of the envelope occurs more fequently leading ultimately to a random spiking type of behaviour.

The dynamics on the upper branch at the same pump intensity values tends to be much more controlled as shown in the right hand side of the figure. Recall that here we are monitoring the peak amplitude of the solitary wave associated with the upper branch solution. It is clear from this figure that one needs to apply larger external stresses to destabilise this upper branch solution. However as shown in the figure the period 2 wavetrain eventually becomes modulated at $|a(0)|^2 = .30$.

The time records in Figure 24 establish that the dynamics of the map is fundamentally altered from that of a plane wave even though we deliberately chose physical parameter values to achieve a quasi plane wave limit. Choosing even larger values for the Fresnel number cannot change the situation as

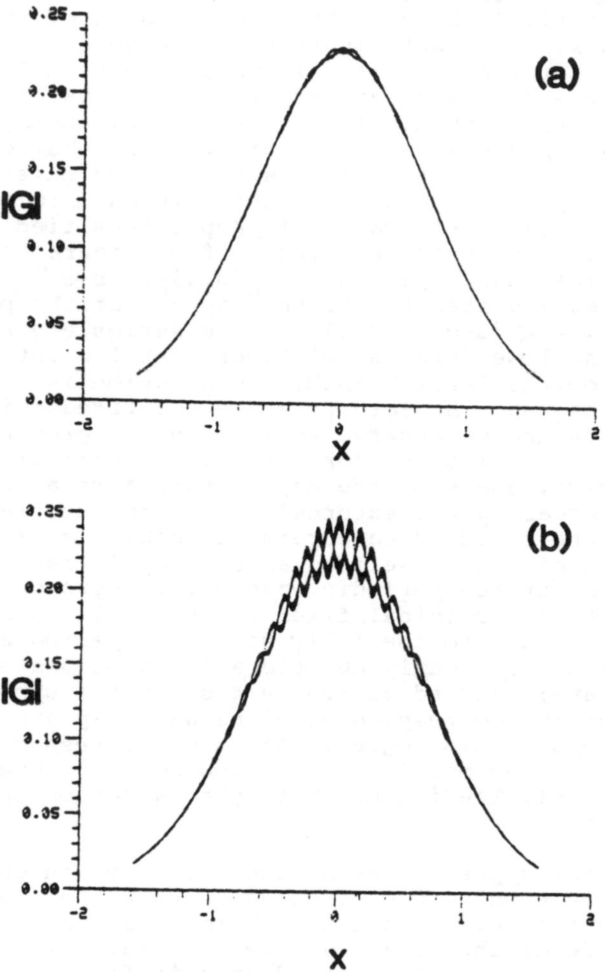

Figure 25. Modulational instability growth on the smooth weakly unstable fixed points at $|a(0)|^2 = .14$ on the lower branch. The corresponding time record is shown at the bottom left of Figure 24. (a) Ten successive outputs starting at n = 140. (b) Ten successive outputs starting at n = 180.

we shall see in the following section. Some insight into the nature of this instability may be gleaned from a close examination of representative transverse profiles just beyond the period doubling bifurcation point. From Figure 25 we see that the instability is purely spatial in origin with the weakly unstable initially smooth fixed point developing high frequency transverse spatial structure. The initial stages of the instability growth are shown in this figure and the period 2 nature of the oscillation is reflected in the alternating switching back and forth between two out of phase modulated envelopes. These high frequency spatial waves continue to grow and eventually saturate, the final stable period 2 state corresponding to an alternation between two fixed spatial shapes. We remark here that the transverse spatial wavelength is a function of the Fresnel number F and can be accurately predicted from the analysis of the next section.

The collapse of the periodic wavetrain towards the original weakly unstable smooth fixed point, alluded to in the discussion of the time records of Figure 24, is clearly illustrated by the sequence of transverse profiles shown in Figure 26 which represent snapshots from the chaotic time evolution at $|a(0)|^2 = .18$ on the lower hysteresis branch in the former figure. The pictures shown illustrate important developments in the dynamic evolution. The first picture shows one of the pair of profiles approaching saturation in the periodic regime. Notice that the amplitudes of the rapid spatial modulations tend to fall off rather rapidly towards the wings. The kinks at the base of each solitary wave are not numerical artifacts but rather are responsible for the impending change in the nature of the solitary wavetrain, shortly before its collapse towards the mean. The second picture shows that the two low amplitude solitary waves on the wings (indicated by the arrows) have split into pairs so that, in effect, two new waves have been introduced within the finite aperture. The final picture shows that most of the rapid spatial modulation has disappeared and the profile has returned close to the original smooth fixed point. The process is now repeated with the exponential growth from the mean. These numerics provide clear evidence for the fact that the modulation instability occurring in the infinite dimensional system can be described as a continual random energy exchange between a finite number of competing nonlinear modes. Our analysis in the following section will lead us to identify those nonlinear modes and isolate the mechanism for instability.

We end this section on numerics by returning to the upper branch instabilities and try to understand these in the light of the results just presented for the lower branch. Figure 27 shows the large amplitude soliton sitting on the broad (lower branch) shelf at $|a(0)|^2 = .16$ at which value it is a stable fixed point. Immediately below this picture is a blow-up of the shelf itself and as expected it is directly superimposable on the lower branch (unstable) fixed point in Figure 25. There is a significant difference between this latter picture and Figure 25 however. Whereas, in Figure 25

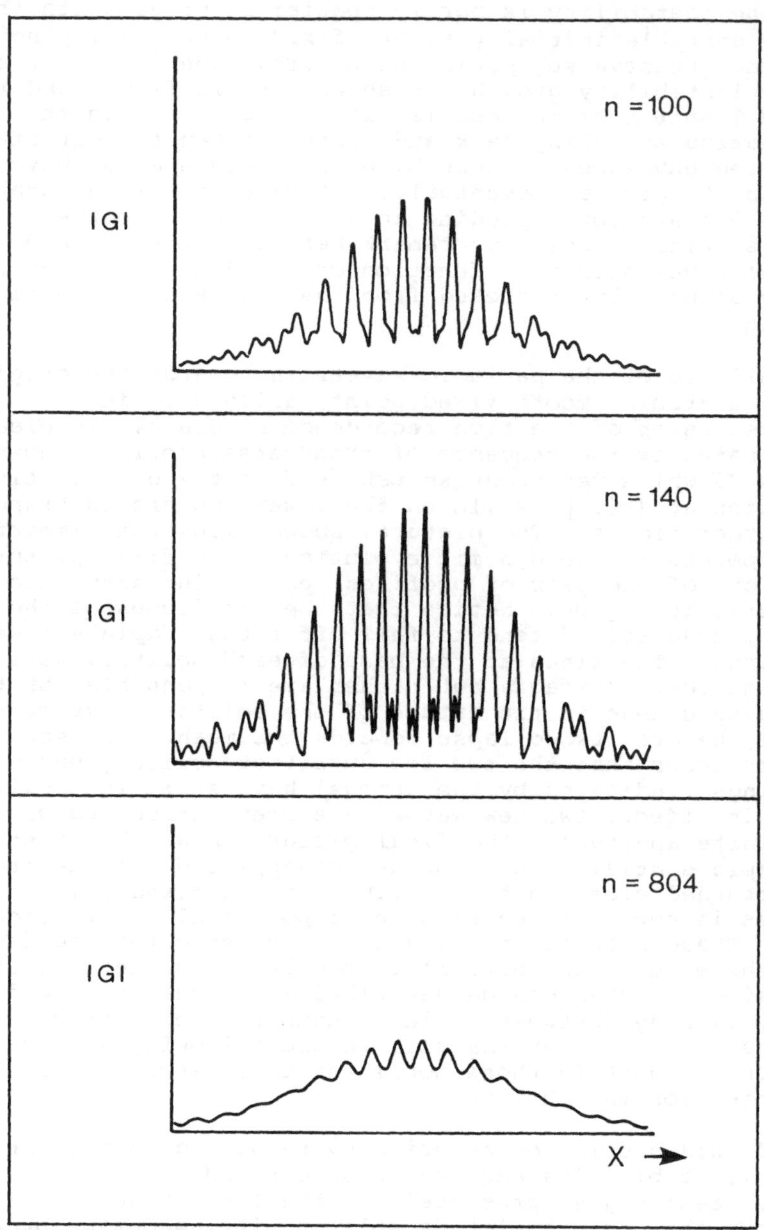

Figure 26. Output at n = 100 shows the growth of the spatial modulation from the initially smooth envelope. At n = 140 the waves at the nodes have split into pairs introducing a further two waves within the finite aperture and at n = 801 the saturated wavetrain has already collapsed towards the smooth fixed point and the modulational growth repeats.

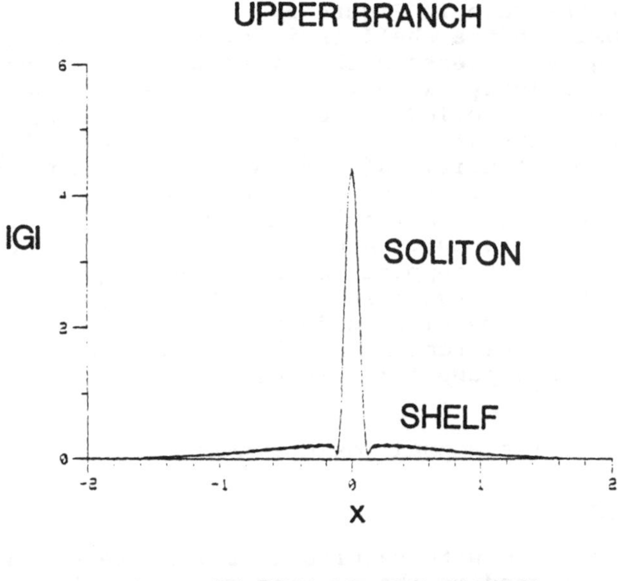

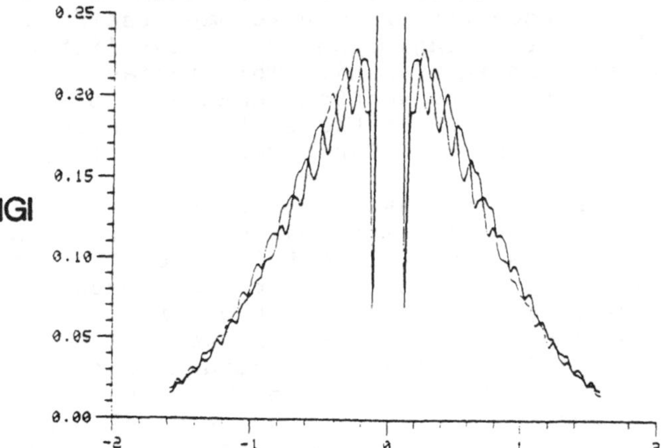

Figure 27. This picture shows the upper hysteresis branch large amplitude soliton sitting on a broad shelf in (a) and the blow-up of the shelf in (b). The shelf is directly superimposable on the smooth lower branch unstable fixed point in Figure 25.

the transverse spatial modulation continues to grow and develop into large amplitude saturated solitary wavetrains, this is not the case in Figure 27. In the latter case the initial growth on the shelf is interrupted at an early stage and this figure represents the final stable asymptotic state. Rather interestingly, we now have a situation where the output exhibits a period 2 oscillation on the shelf but is a stable fixed point at centre (Moloney, 1986b). As the external pump intensity $|a(0)|^2$ is increased further, the modulational instability on the shelf (which is, in effect, the lower branch instability) wins out and literally shakes the large soliton about. A numerical experiment was carried out to lend further support to the above argument. We took a chaotic output at large pump intensity $|a(0)|^2 = .34$ and removed the shelf by closing down an aperture. The end result was that the remaining soliton was stable over a wide range of external pump intensities.

6. ORIGIN OF THE MODULATIONAL INSTABILITIES

6.1 Analysis

This section deals with results from some recent analysis which was motivated by the earlier discovery by McLaughlin et al. (1985) that the plane wave map is more unstable to perturbations with transverse stucture than to plane wave perturbations. Hence the plane wave map itself, while mathematically interesting, appears to bear little relevance to the physical problem at hand. This instability, as we have seen in the previous section, changes the whole character of the route which the system takes in going from a simple to a chaotic state. Furthermore, the instability is universal in character and should have ramifications for a large class of nonlinear wave problems whose dynamics are described by envelope equations which are forced and damped at regular intervals. We will not dwell on this latter point here but instead direct the reader to the recent review article by Aceves et al. (1986). In order to prepare the groundwork for the following analysis we will now briefly outline the stability analysis of McLaughlin et al. (1985). Further details may be found in the above review article, in Moloney 1986b and, of course, in the original paper. The original analysis was motivated by the numerical results for gaussian pump beams just presented in the previous section.

The starting point of the analysis is to take the plane wave fixed point g determined from eqn. (3) and investigate its stability with respect to perturbations with transverse structure. In other words we set

$$G_n(\underline{x},z) = (\, g + y_n(\underline{x},z)\, e^{\frac{i}{2}\, p\, N(gg^*)\underline{x} + i\arg g} \tag{34}$$

in the full map (eqns. (1) and (2)). The perturbation

$y_n(\underline{x},z)$ has the following general form

$$y_n(\underline{x},z) = e^{i\nu z}\,(a_n e^{i\underline{k}\cdot\underline{x}} + b_n e^{-i\underline{k}\cdot\underline{x}}) + \qquad (35)$$
$$e^{-i\nu z}\,(c_n e^{i\underline{k}\cdot\underline{x}} + d_n e^{-i\underline{k}\cdot\underline{x}})$$

for $4\nu^2 = \gamma^2 k^4 - 2p\gamma\, N'(I)Ik^2 \quad 0$, $(k^2 = k_x{}^2 + k_y{}^2$, $I = gg^*)$ and

$$y_n(\underline{x},z) = e^{\sigma z}\,(a_n e^{i\underline{k}\cdot\underline{x}} + b_n e^{i\underline{k}\cdot\underline{x}}) + \qquad (36)$$
$$e^{-\sigma z}\,(c_n e^{i\underline{k}\cdot\underline{x}} + d_n e^{-i\underline{k}\cdot\underline{x}})$$

for $4\sigma^2 = 2p\gamma N'(I)I\,k^2 - \gamma^2\,k^4 \quad 0$.

The second form of the solution is realised by an analytic continuation $\nu = -i\sigma$ and, if $\sigma \quad 0$, implies exponential growth of the perturbation on propagation through the nonlinear medium. This latter instability is reminiscent of the Benjamin-Feir or modulation instability which is so widespread in physics. The first form for the transverse perturbation is clearly purely oscillatory and hence does not represent an instability on propagation. Recall however that $y_n(\underline{x},z)$ represents just a part of the stability problem and this must, in turn, be substituted into the cavity boundary condition to yield the linearisation of infinite dimensional map. The resultant linear map becomes

$$y_{n+1}(x,o) = Re^{i\phi_0}y_n(\underline{x},L) \qquad (37)$$

Substituting in the appropriate value of $y_{n+1}(\underline{x},o)$ and $y_n(\underline{x},L)$ into this map (we seek an eigenvalue crossing the unit circle through -1) we end up with an expression for the potentially unstable root ρ, i.e.

$$\rho/R = b + \sqrt{b^2 - 1}$$

where

$$b(\mu,\tau) = \cos(\psi+\mu)\cos\nu + \frac{(\tau-2\mu)}{2\nu}\,\sin(\psi+\mu)\sin\nu \qquad (38)$$

with $\psi = \pi + \phi_0 - \tfrac{1}{2}p$, $2\nu = (\tau^2-4\mu\tau)^{\frac{1}{2}}$ where $\tau = \gamma k^2$ and $\mu = pI$. This final stability formula is derived by assuming a Kerr nonlinearity ($N(I) = -1 + 2I$) as we are concerned with the stability of low amplitude solutions on the lower hysteresis branch. The stability formula is widely applicable, yielding unstable plane wave fixed points for both signs of the nonlinear coefficient p ($p < 0$ corresponds to a self-defocussing nonlinearity). For a more complete discussion the reader is referred to the original article of McLaughlin et al. (1983) and to two recent articles (Moloney, 1986a,1986b). As remarked earlier, there exists a parameter region where the pertubation term $y_n(\underline{x},z)$ is purely oscillatory throughout the nonlinear medium. However, rather surprisingly, the map itself is unstable to the growth of modulational transverse structures over such parameter regions. The next subsection will specifically address this interesting case and we will find that it corresponds to the case investigated numerically in the last section.

Again, we will confine our study to a single transverse spatial dimension although the formulae presented here suffer from no such restriction.

6.2 Spatial sideband instability

This section presents some very recent results which throw further light on the nature of the instability discovered numerically in section 5. The numerical experiments to be described here establish, unequivocally, that the modulation instability involves the participation of just a finite number of nonlinear spatial modes. Moreover it suggests a natural truncation of the infinite dimensional map to one involving just three modes. The work to be described here, while preliminary (Adachihara) is particularly enlightening. We argue that we can achieve considerable insight into the nature of the gaussian beam instabilities of section 5 by studying numerically the instabilities of an initially plane wave, subject to periodic boundary conditions. This approach is justified, a posteri, by the remarkable similarities to the behaviour observed in the numerics of section 5.
An obvious advantage of this approach is that we can use the analytic stability formula for $b(\mu,\tau)$ to provide us with quantitative information on the dynamically relevant modes, at least close to the instability threshold. The parameters are chosen to be identical to those used in section 5 with the exception of the critical external pump intensity $|a(0)|^2$, which is now different as we are dealing with a plane wave. The other available degree of freedom is the transverse extent of the plane wave ($-L < x < L$), the choice of which is crucial in order to get an instability at all. Figure 28 shows a plot of $b(\mu,\tau)$ vs $k(k \equiv k_x)$ for the following set of parameters $p = 6$, $F = 33.3$, $\phi_0 = .4$, $R = .9$ and $|a|^2 = .110$. The horizontal dashed line represents the critical value of b at $\rho = -1$ $\left[(b_c = \frac{1}{2}(R+1/R)\right]$. There is clearly a continuous band of transverse spatial wavenumbers k which lie above critical. All wavenumbers within this band can grow exponentially but the one experiencing the maximum growth rate (maximum of $b(\mu,\tau)$) will ultimately win out. Notice that a plane wave ($k = 0$) is linearly damped and that the entire band lies in the so-called oscillatory region.

The situation just described refers to an infinite plane wave. Imagine now that we confine this plane wave in the transverse dimension by introducing, say, a finite aperture of width 2L ($-L < x < L$). The net effect of this aperture (periodic boundary condition) is that the spectrum of wavenumbers is now discrete rather than continuous. As we

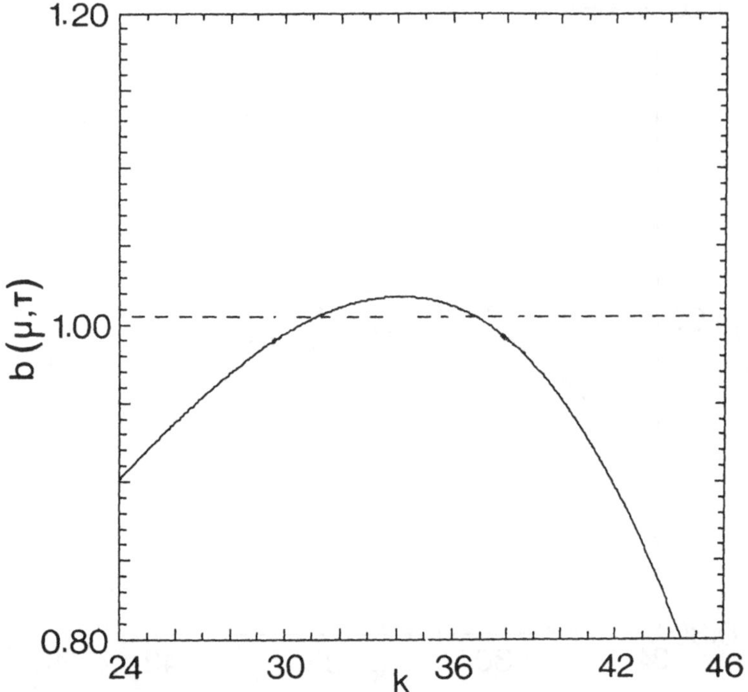

Figure 28. Instability curve $b(\mu,\tau)$ for an infinite plane wave. The continuous band of wavenumbers lying above the critical line b_c (dashed) are potentially unstable to a modulational growth. The wavelength ultimately chosen corresponds to the maximum in $b(\mu,\tau)$. Note that the entire band lies in the oscillatory region and that the plane wave (k=0) is linearly damped for the present set of parameter values.

gradually close down the aperture we increase the separation between adjacent discrete wavenumbers and eventually can reach a situation where none of the wavenumbers lie within the unstable band shown in Figure 28. In this latter circumstance there is no instability. The situation just described is the precise analogue of changing, say, the focal spot size of a gaussian beam (i.e. changing the Fresnel number F). We can thus imagine the present plane wave problem with periodic boundary conditions to be a first approximation to a gaussian beam.

Consider the following problem. We want to repeat the computations which led to Figures 24 and 25 for the gaussian beam but instead use an apertured plane wave. Referring to these figures we observe that the modulations on the gaussian beam span a range $-1.6 < x < 1.6$. Using this range for the aperture width and an appropriate value for the plane wave fixed point amplitude we get the situation shown in Figure

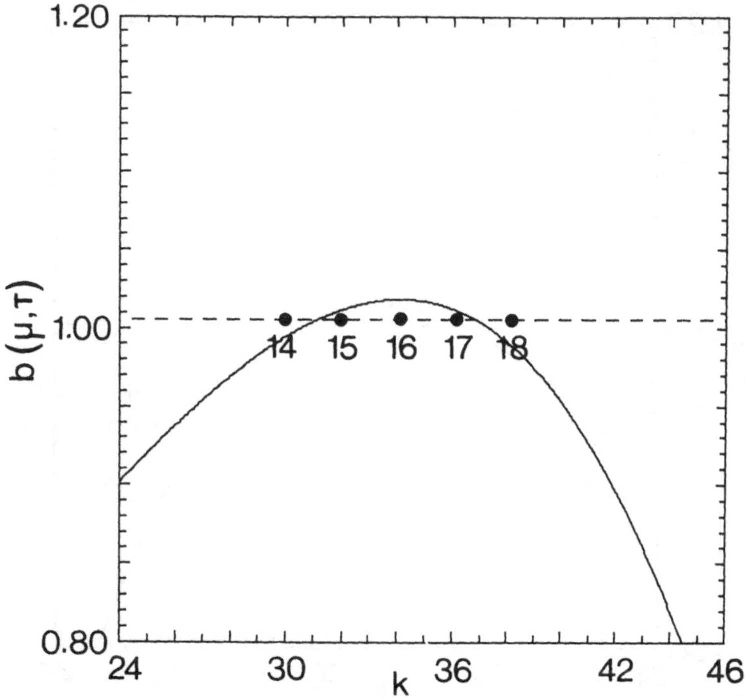

Figure 29. Instability curve $b(\mu,\tau)$ for an apertured plane wave ($|x| > 1.6$). This picture is identical to Figure 28 with the exception that only discrete k vaues are allowed. The location of the even (k_{14}, k_{16}, k_{18}) and odd (k_{15}, k_{17}) modes with respect to the band $b(\mu,\tau) > b_c$ are indicated.

29. Notice from this figure that a 16 wave modulation lies right under the most unstable region of the band. Two odd wave modulations (15 and 17) also lie in the band and the even pair 14 and 18 just outside. This figure tells us that the apertured plane wave should be unstable to the growth of a 16 wave transverse spatial modulation. Figure 29 now sets the scene for the following careful numerical study. In all of the numerics described here the L^2-norm and Hamiltonian were monitored throughout the nonlinear medium propagation. An error of at most, .05% was observed in this latter quantity.

Starting from the plane wave fixed point, derived from an external pump intensity, $|a|^2 = .110$, as initial data to the infinite dimensional map (eqn. (4) and (5)), we introduce a small even perturbation of the form $G_n(x,o) = (|g| + \varepsilon \cos k_{16}x)e^{i\theta + i \arg}$ with the perturbation amplitude $\varepsilon = .001$ and the corresponding fixed point amplitude $|g| = .18711$. This is necessary to seed the instability as the plane wave perturbation itself is linearly damped ($k = 0$) (see Figure 29). As the perturbation is even we can ignore the odd modes

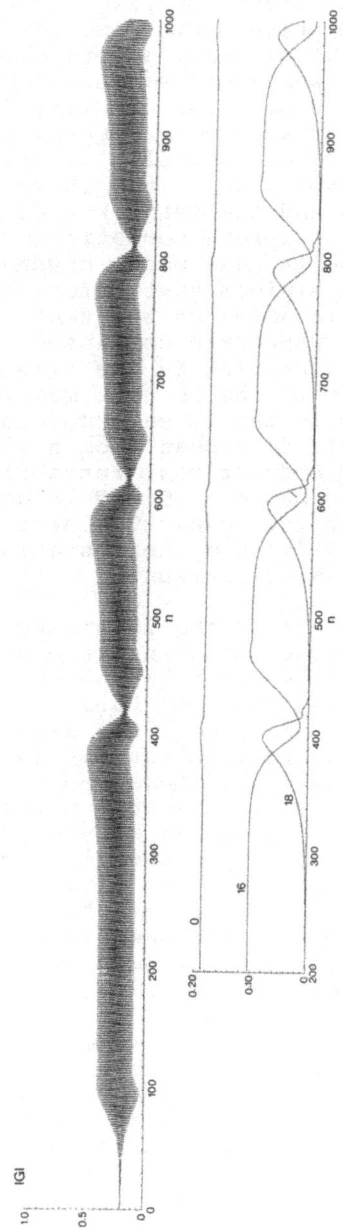

Figure 30. (a) Time record at the centre of the aperture. (b) Time record of the corresponding Fourier amplitudes of the mean (k = 0), the 16- and 18- wave modulations.

(15 and 17) so that the nearest modes that can most strongly interact with the 16 wave mode are the even pair (14 and 18). Figure 30(a) shows the time record at the centre (x = 0) showing the initial destabilisation of the plane wave fixed point with saturation to a period 2 cycle beyond n = 120. Notice the remarkable similarity with the time record for the centre of the lower branch profile in Figure 24. In fact the saturated period 2 state (involving 16 waves) is marginally stable; convergence of every second output to seven significant digits was achieved at n = 200. At this point (n = 200) a small 18 wave perturbation is introduced as a seed. This now starts to grow competing with the 16 wave state. Figure 30(b) dramatically illustrates this phenomenon showing the time evolution of the 16 wave and 18 wave spatial Fourier amplitudes and the mean (k = 0) starting from n = 200. Notice the strong competition between the growing 18 wave and falling 16 wave modes beyond n = 340. Both wave amplitudes reach a minimum just before n = 420 and if we refer back to Figure 30(a) we see that this is precisely the point at which the wavetrain collapses towards the mean. There is also a slight kink in the time record of the mean in this collapse region. The 16 wave modulation starts to grow again and saturate as the 18 wave modulation dies away. The whole process starts to repeat from n > 460. We conclude therefore that the modulational instability involves an endless competition between 16 and 18 wave modulations interacting through the mean. The next sequence of pictures of transverse profiles and their spatial Fourier transforms further support this conclusion.

Figure 31(a) shows one of the saturated 16 wave period 2 states at n = 200 together with its spatial Fourier transform. As remarked earlier the state has converged to seven significant digits. The Fourier transform (Fig. 31(b)) shows that the state is not a pure sinusoid but has many higher harmonics, suggesting that it is a 16 wavetrain of solitary waves. Figure 32 shows the solitary wave train and its spatial Fourier transform at n = 202 immediately after the 18 wave seed is introduced at n = 201. The spectrum now shows a strong 16 wave amplitude plus lower amplitude 18, 2 and 14 waves. These latter two waves arise directly as a consequence of a nonlinear interaction between k_{18} and k_{16} and k_{18} and k_{32} (the second harmonic of k_{16}). Notice that the k_{18}, k_2 and k_{14} are still much smaller than k_{16}; the vertical axis is on a logarithmic scale.

Having established that the 18 wave seed introduces, through nonlinear interaction, two further growing spatial modulations corresponding to 2 and 14 waves, we now show how these additional modulations are responsible for the collapse of the 16 solitary wavetrain. Figure 33 shows a sequence of eight pictures as we approach and pass through the first collapse region in Figure 30(a). The top two pictures show the growing influence of k_2 and k_{14} at n = 300 and 340. Notice the increasing similarity of the slow envelope (k_2) to the envelope of the solitary wavetrain for the gaussian pump

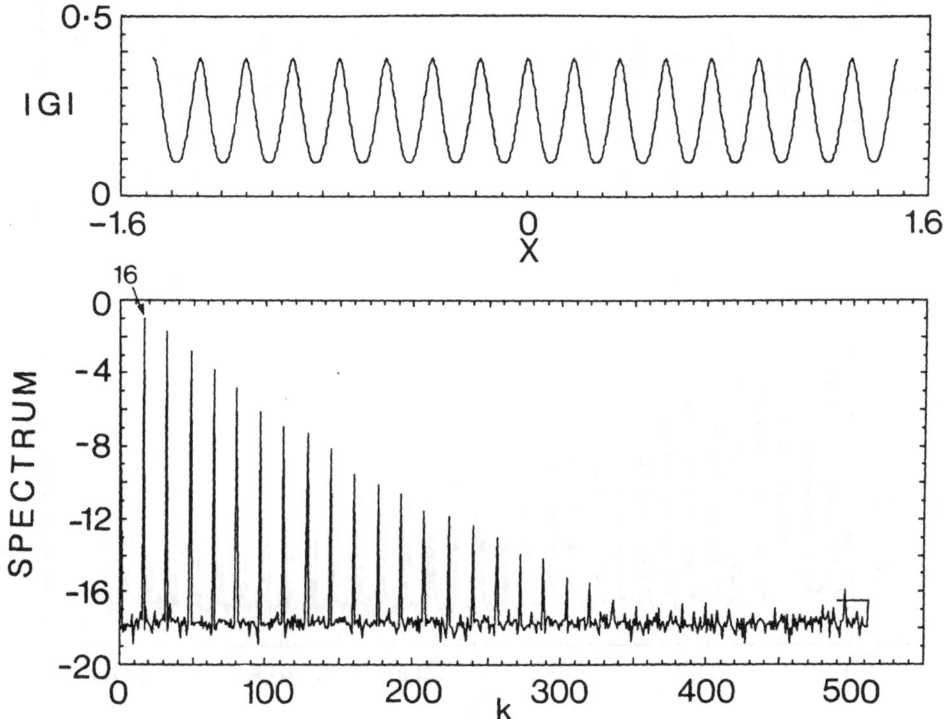

Figure 31. (a) Transverse spatial profile showing one of the saturated 16 wave period 2 states at n = 200. (b) The Fourier spectrum showing the 16 wave peak and its higher harmonics.

beam in Figure 25. The next fundamental change in the solitary wavetrain occurs at n = 350 and 360. The next two pictures show that the waves start to split but the only two that succeeed are two symmetrically disposed low amplitude waves at the nodes of the envelope. We now have the interesting situation that the number of solitary waves within the box has increased by two to 18 solitary waves. At n = 390 this new 18 wavetrain is still growing but it never sheds the slow (k_2) envelope and it starts to fall off in amplitude as shown at n = 410. Finally the collapse is complete by n = 420 and the solution has approached close to the mean (the initial k = 0 data). As we are now back close to the instability boundary, figure 29 is again relevant and we expect the 16 wavetrain to experience maximum growth. This is the case at n = 420 and 430 where 16 waves again tile the box and the whole process repeats. The process here parallels the way in which extra rolls are introduced in one dimensional convection patterns (Aceves et al., 1986).

One can conclude from the preceding discussion that the modulational chaos arises as a consequence of a nonlinear

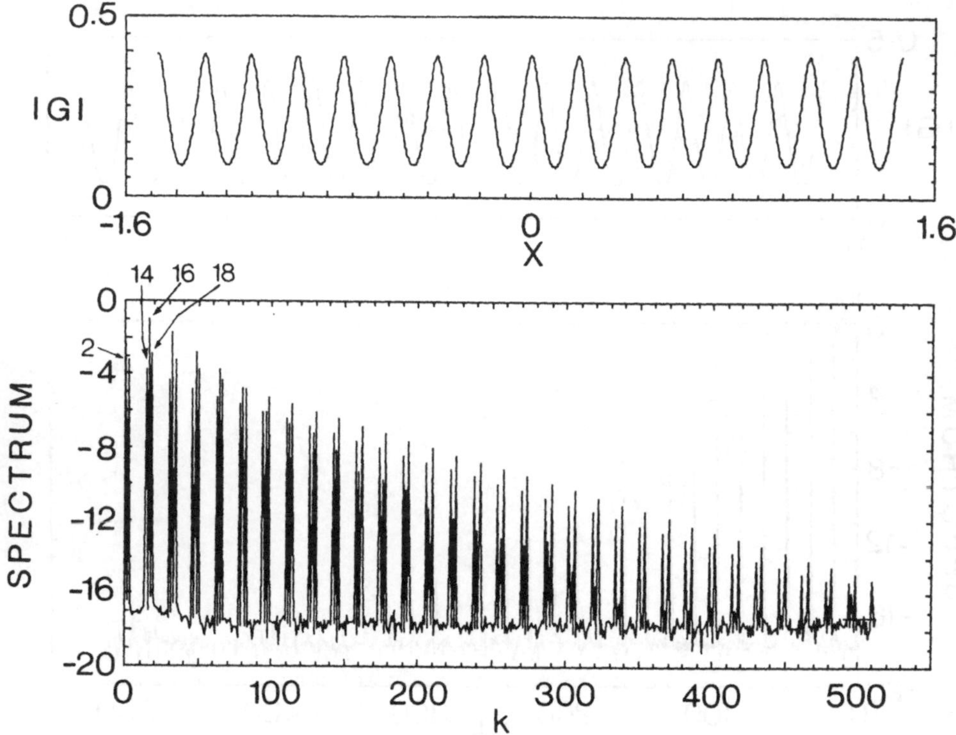

Figure 32. (a) Transverse spatial profile at n = 201 immediately after the introduction of the 18 wave seed. (b) The corresponding Fourier spectrum showing the presence of k = 14, 16, 18 and 2 wavenumber peaks.

interaction between three natural modes. The initial state (k = 0), representing the mean, is weakly unstable to a 16 wave growth lying at the centre of the unstable band. This fully saturated solitary wavetrain (k_{16}) tiles the box defined by the finite aperture ($-L < x < L$) and is again weakly unstable to an 18 wave perturbation. Nonlinear interaction between the fundamental waves and/or their harmonics leads to a slow envelope modulation of the initial 16 solitary wavetrain. The individual waves try to split by introducing kinks at their bases but only the lowest amplitude ones, closest to the mean, succeed. An 18 wave modulation starts to grow but cannot tile the finite box and suddenly collapses back towards the weakly unstable mean, thereby initiating the whole process again. The crucial point is that the three natural participating nonlinear modes, namely, the mean and the 16 and 18 wave modulations, are each unstable. One may then envisage a trajectory in phase space which passes close to a heteroclinic orbit connecting these three unstable fixed points, the distance of closest approach being completely random. Hence the modulational chaos. The next step is to test these

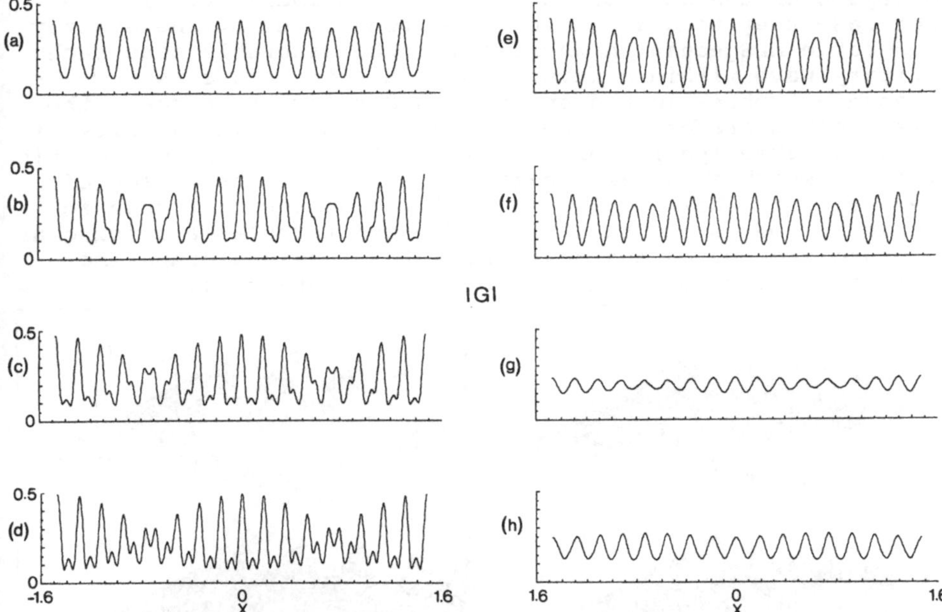

Figure 33. Sequence of eight snapshots from the dynamic evolution showing the development of the slow modulation at k_2 (n = 300 and 340) of the 16 wave state, the splitting effect (n = 350 and 360), growth and beginning of the collapse of the 18 wave state towards the mean (n = 390-420) and repeat of the 16 wave modulation growth from the mean (n = 420-430).

hypothesis by linearising about each of the fixed points (the mean is already done) in order to determine the local directions of their stable and unstable manifolds. This work is currently underway (Adachihara).

7. CONCLUSIONS

We hope that the analysis presented in this article has convinced the reader of the power of a blend of a computer and analytic approach to the study low level turbulence in complex physical systems. No a priori assumptions need be made to reduce a complex mathematical problem to a manageable one. Rather the computer generated solutions directly suggest the natural modes for the problem. The class of physical problems to which the ideas and techniques developed in this article may apply should be very wide indeed.
The natural modes of the infinite dimensional map describing the propagation of a laser beam in a passive optical resonator, have been identified with the solitary wave/soliton solutions of a nonlinear PDE of the nonlinear

Schrodinger type. The nature of these fixed points is seen to depend on the form of the optical nonlinearity. In the Kerr case they tend to mimic N-soliton states of the exactly integrable NLS equation whereas for the saturable case each solitary wave within a wavetrain, tends to behave as an independent phase locked entity. Under further stress, the fixed points are destablised and a modulation instability of the transverse spatial profile leads to a new type of chaotic dynamics which is purely spatial in origin. The motion of the system is shown to lie on a finite dimensional chaotic attractor and is due to a form of phase turbulence arising as a consequence of an endless competition between competing nonlinear modes confined within a finite aperture.

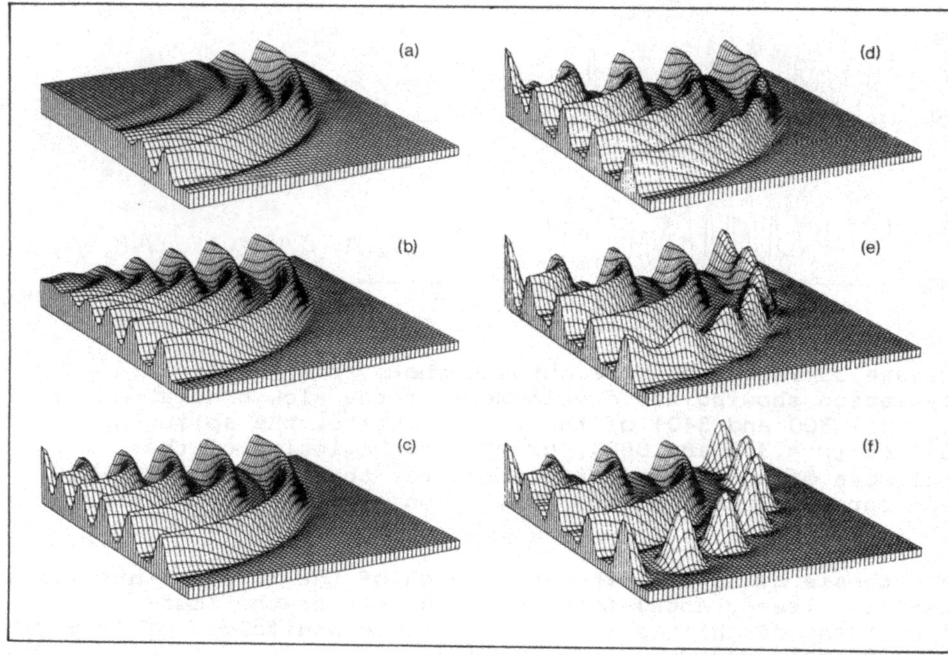

Figure 34. Dynamical switching and solitary wave evolution across the two transverse dimensional beam profile (one quadrant of the profile is displayed). Parameters used to generate these pictures are the same as those used in Figures 6-8. A secondary instability appears on the outermost ring at n = 420 (see (d)) and develops into a ring of filaments (see (f)).

Directions for future work on this map are clear. The obvious step of extending the study to two transverse dimensions is already well underway with some interesting new nonlinear dynamic features already evident. Specifically, we have investigated the dynamical switching of the two transverse dimensional profile to the upper hysteresis state with parameter values corresponding to the one transverse dimensional switching of Figures 6-8. Figure 34 shows part of this switching process for a saturable nonlinearity; the cubic nonlinearity leads to filamentation in two transverse dimensions (Moloney, 1985, 1984b). The cylindrical symmetry of the problem is preserved during the early phase of the evolution suggesting that much computational effort could be saved by reformulating the problem in cylindrical coordinates and solving the resulting one dimensional radial PDE. In fact the persistent oscillations of the central part of the beam (not present in one dimension) mimics the behaviour of solitary wave solutions to this latter equation. However, under further iteration ($n > 400$) the outer ring begins to break into filamentary structures. These filaments tend to be stabilised by saturation effects but they do show evidence for an amplitude turbulence by disappearing and reappearing in random fashion at different parts of the physical domain. An open question, at the moment, is what happens in this system under increased external stress? The modulational instability results of section 6 are directly relevant to this situation. The degeneracy in the two dimensional transverse wave vector $\underline{k}$ (only its amplitude is chosen by the instability), suggests the possibility of complex two dimensional spatial pattern formation.

ACKNOWLEDGEMENTS

Two of the authors (JVM and ACN) acknowledge a NATO research grant which maintains the collaborative research effort and made the writing of this manuscript possible. Much of the work described here was supported by research grants from the US Air Force Office for Scientific Research, the Office of Naval Research, the U.S. Army Research Office and the National Science Foundation.

REFERENCES

Aceves, A., Adachihara, H., Jones, C., Lerman, J.C., McLaughlin, D.W., Moloney, J.V. and Newell, A.C., 1986, Physica, 18D, 85.

Adachihara, H., McLaughlin, D.W., Moloney, J.V. and Newell, A.C., J. Math. Phys. (submitted).

Adachihara, H. (unpublished).

Bowden, C.M., Ciftan, M. and Robl, H.G. (ed.), 1981, Optical Bistability 2 (New York: Plenum).

Bowden, C.M., Gibbs, H.M. and McCall, S.L. (ed.), 1984, Optical Bistability 2 (New York: Plenum).

Gibbs, H.M., 1985, Optical Bistability: Controlling Light with Light (London, New York: Academic Press).

Hammel, S., Jones, C.K.R.T. and Moloney, J.V., 1985, J. Opt. Soc. Am. B, 2, 552.

Ikeda, K., 1979, Opt. Commun., 30, 256.

Ikeda, K. and Akimoto, O., 1980, Phys. Rev. Lett., 45, 707.

Mandel, P. and Kapral, R., 1983, Opt. Commun., 47, 151.

McLaughlin, D.W., Moloney, J.V. and Newell, A.C., 1983, Phys. Rev. Lett., 51, 75.

McLaughlin, D.W., Moloney, J.V. and Newell, A.C., 1985, Phys. Rev. Lett., 55, 168.
Moloney, J.V., Hopf, F.A. and Gibbs, H.M., 1982, Phys. Rev. A, 25, 3442.
Moloney, J.V., 1984a, Opt. Commun., 48, 435.
Moloney, J.V., 1984b, Phil. Trans. Roy. Soc., A313, 429.
Moloney, J.V., 1984c, Phys. Rev. Lett., 53, 556.
Moloney, J.V., 1985, J. Quant. Electron. IEEE, QE-21, 1393.
Moloney, J.V., 1986a, "Plane Wave and Modulational Instabilities in Passive Optical Resonators", in Nonlinear Phenomena and Chaos, ed. S. Sarkar (Adam Hilger, Bristol, Boston), 214.
Moloney, J.V., 1986b, Phys. Rev. A, 33, 4061.
Newell, A.C., 1985, "Solitons in Mathematics", (SIAM Philadelphia).
Scott, A.C., Chu, F.Y.F., McLaughlin, D.W., 1973, Proc. IEEE, 61, 1443.
Shuster, H.G., 1984, "Deterministic Chaos" (Physik-Verlag).

THE ARITHMETIC OF CHAOS

F VIVALDI

ABSTRACT

The Anosov diffeomorphisms of the two dimensional torus are the most chaotic area-preserving maps. The study of their periodic orbits is transformed here into a problem of arithmetic in certain real quadratic fields. This procedure allows one to classify and construct all periodic orbits, and reveals the arithmetical nature of the laws which govern chaotic motions.

1. INTRODUCTION

This paper is about periodic orbits of the most chaotic area-preserving maps: the Anosov diffeomorphisms of the 2-dimensional torus (for a review, see Franks, 1970). These systems display chaos in its purest form, which is much simpler than that of generic canonical maps. Indeed Anosov diffeomorphisms do not have regular orbits (i.e. divided phase space), for they are everywhere hyperbolic. For this reason the periodic orbits of these systems can be "completely" understood (as much as chaos can be understood). The situation here is somewhat similar to that of integrable systems, which can be solved because all their orbits are regular. However, orbits of integrable systems are studied mainly with the tools of analysis, those of Anosov systems with arithmetic.

Manning (1975) proved a result which is crucial for our purpose: every Anosov diffeomorphism of the n-dimensional torus $\mathbf{T}^n$ is topologically conjugate to a linear Anosov diffeomorphism. Thus we can represent Anosov diffeomorphisms (up to topological conjugacy) by means of elements of $SL_n(\mathbf{Z})$, i.e. n-dimensional matrices with integral entries, determinant equal to one, and real eigenvalues (for hyperbolicity). The periodic orbits of these matrices have rational coordinates, and vice-versa (this can be verified almost effortlessly). It follows that every point of a periodic orbit must belong to some n-dimensional lattice on the torus, namely the collection of the points $(m_1/g, \ldots, m_n/g)$, for some fixed integer g and integers m_i, $0 \le m_i < g$, $i=1, \ldots, n$. Such lattices are clearly invariant under the action of integral matrices. So the study of periodic orbits can be reduced to a problem in integers. The goal here is to determine the number of orbits which exist on each invariant lattice, along with their periods and initial conditions (see §3). One would also like to characterize the lattice to which all orbits of a given period belong (§4).

In order to transform the dynamics of periodic orbits into a problem of arithmetic, we shall use a classical result of Latimer and MacDuffee (1933) and Taussky (1951), where a one-to-one correspondence was found between conjugacy classes in $SL_n(\mathbf{Z})$ (defined by the equivalence relation: $A \sim B$ if X exists, such that $A=XBX^{-1}$; $A,B,X \in SL_n(\mathbf{Z})$) and the group of ideal classes in certain rings of algebraic integers (see Taussky (1978), for a review). The idea is to identify the invariant lattices mentioned above with ideals, thereby providing them with an arithmetical structure (i.e. a multiplication). This correspondence will be illustrated in §2, for the 2-dimensional case.

This paper is a report of work which has been published elsewhere (Percival and Vivaldi, 1986 [PV]). The main results, which are proved in [PV], are listed here as propositions A1-A8. The reader is assumed (here, but not in [PV]) to be familiar with the arithmetic in algebraic number fields. We shall be entirely concerned with the two-dimensional case, which corresponds to quadratic fields. For a classical reference, see Hecke (1923);

for quadratic fields, see Cohn (1962).

2. THE CORRESPONDENCE PROCEDURE

The object of our study is the periodic orbits of matrices M of $SL_2(\mathbf{Z})$. Rather than starting from M, we start from its characteristic polynomial, which is a monic polynomial over $\mathbf{Z}$, of degree 2: $f(x)=x^2-kx+1$, $k\in\mathbf{Z}$, $|k|>2$ (k is the trace of M). Its roots are quadratic integers in the real quadratic field $\mathbf{Q}(\sqrt{D})$, D being the square-free kernel of $k^2-4>0$. The ring R of all algebraic integers in $\mathbf{Q}(\sqrt{D})$ has discriminant d and $\mathbf{Z}$-basis $[1,\omega]$, where

$$\begin{array}{lll} d = 4D, & \omega = \sqrt{D}, & D \not\equiv 1 (\mathrm{mod}\ 4); \\ d = D, & \omega = (1+\sqrt{D})/2, & D \equiv 1 (\mathrm{mod}\ 4), \end{array} \tag{2.1}$$

i.e. every $z\in R$ can be represented uniquely as $z=x+y\omega$, $x,y\in\mathbf{Z}$.

Our purpose is to classify all matrices whose characteristic polynomial is $f(x)$. Let λ be a root of $f(x)$ (the largest one, say). Then the norm of λ is equal to 1 (the norm $N(z)$ of a quadratic integer z is the rational integer zz', z' being the algebraic conjugate of z), which means that λ is a *unit*, i.e. a divisor of all integers in R. This property plays a central role in the theory, being the the arithmetical equivalent to area-preservation. Let now $I=[\omega_1,\omega_2]$ be an ideal in R. Since $\lambda I=I$ (λ is a unit), the eigenvalue equation

$$\lambda\begin{pmatrix}\omega_1\\ \omega_2\end{pmatrix} = \begin{pmatrix}a & c\\ b & d\end{pmatrix}\begin{pmatrix}\omega_1\\ \omega_2\end{pmatrix}, \qquad M = \begin{pmatrix}a & b\\ c & d\end{pmatrix} \tag{2.2}$$

defines a strictly unimodular matrix M (Det(M)=+1), the determinant of M being equal to the norm of λ. The ideal class to which I belongs (namely, the set of ideals J for which $zI=wJ$, for some $z,w\in R$) determines uniquely a matrix class in $SL_2(\mathbf{Z})$. Indeed a change of basis for I corresponds to conjugacy in $SL_2(\mathbf{Z})$, while if J is in the same class as I, then $J=\alpha I=[\alpha\omega_1,\alpha\omega_2]$, for some $\alpha\in\mathbf{Q}(\sqrt{D})$, and therefore J generates the same matrix M as I.

If h is the class number (the number of ideal classes) of R, Eq. (2.2) allows one to construct h matrix classes in $SL_2(\mathbf{Z})$, all sharing the same eigenvalues. Clearly, all matrices which belong to the same class will have the same orbits, since conjugacy in $SL_2(\mathbf{Z})$ corresponds to canonical transformations. We shall see that the structure of periodic orbits is largely independent of the class, which means that all Anosov diffeomorphisms which are constructed from ideal classes in R have essentially the same dynamics, even if they are not canonically conjugate.

We must remark, however, that the Latimer-MacDuffee-Taussky correspondence identifies matrix classes in $SL_2(\mathbf{Z})$ with ideal classes in the ring (order) $O=[1,\lambda]$, which is in general contained properly in R, and has a larger class number. Ideal classes in O can be constructed from those in R using the notion of ring equivalence (Cohn, 1978, p. 143). Here, for simplicity, we consider only those matrix classes which originate from ideal classes in R, to avoid the computational difficulties related to the non-unique factorization of some ideals in O. Specifically, if $O=[1,\lambda]$, then, for some rational integer f, $O=[1,f\omega]$, with ω given by (2.1). The rational prime divisors of f do not factor uniquely in O, since O is not integrally closed for $f>1$. However, our analysis will still apply to ideal classes in O (and the corresponding Anosov diffeomorphisms) provided one considers only lattices which are relatively prime to f.

We will now transform dynamics into modular arithmetic in R. We begin by choosing a representative ideal I from each ideal class in R, and construct the corresponding matrix M by means of Eq.(2.2). We are interested in the action of M on $\mathbf{Z}^2/g\mathbf{Z}^2$, which represents a g by g lattice on the two dimensional torus. From Eq.(2.2) we see that such action corresponds to multiplication by λ in the R-module I/gI, once we identify (1,0) with ω_1 and (0,1) with ω_2, respectively. The point (x,y) in $\mathbf{Z}^2/g\mathbf{Z}^2$ is now the residue class of $z=x\omega_1+y\omega_2$, modulo the ideal gI. To determine the period of the orbit through z we must find the smallest integer T for which

$$\lambda^T z \equiv z(\text{mod } gI). \tag{2.3}$$

Since $z\in I$, $z=IJ$, for some ideal J, whence $(z,Jg)=I(J,g)$. Eq.(2.3) becomes

$$\lambda^T \equiv 1(\text{mod } (g)/(J,g)), \tag{2.4}$$

which, for fixed λ and g, depends only on (J,g). If $(J,g)=(1)$, the orbit through z will be called a *free orbit*, and an *ideal orbit* otherwise. The possible values of (J,g) are determined uniquely by the ideal factors of (g), which correspond to modules in I/gI. This is because I/gI is isomorphic (as an R-module) to R/gR. This isomorphism sends $z\in R/gR$ to $wz\in I/gI$, w being any (fixed) element of I for which $(w,gI)=I$ (such a w can always be found). It follows that all matrices M which are constructed from ideals in R have the same number of orbits with the same period and on the same lattices, just rearranged differently. Using the above isomorphism one can then construct all orbits for the matrix generated by any ideal I, once those of the matrix generated by $I=(1)=R=[1,\omega]$ are known. For this reason we can and will restrict ourself to this latter case, which is the simplest from the viewpoint of computation.

3. PERIODIC ORBITS ON GIVEN LATTICES

The natural starting point is the case $g=p$, a rational prime. Then the structure of periodic orbits will depend on the factorization of (p) in the quadratic field $\mathbf{Q}(\sqrt{D})$, which is determined by the following rules

$$
\begin{array}{lll}
(p) = (p) & \text{iff } (d/p)=-1 & \text{p is inert} \\
(p) = P_1P_2 & \text{iff } (d/p)=+1 & \text{p splits} \\
(p) = P_1^2 & \text{iff } (d/p)=0 & \text{p ramifies}
\end{array}
\qquad (3.1)
$$

Here d is the discriminant of $\mathbf{Q}(\sqrt{D})$ (see (2.1)) and (d/p) is the Kroenecker symbol (Cohn, 1962, Ch. II, §6). Accordingly, we have three different possible orbit behaviours, as detailed below (the fixed point z=0 is ignored).

A1 If $(d/p)=-1$, (p) has no factors. All orbits are free and have the same period T, which is a divisor of p+1. If $T=(p+1)/m$, then there are $m(p-1)$ free orbits.

A2 If $(d/p)=1$, $(p)=P_1P_2$ has two ideal factors. All orbits have the same period T, which divides $p-1$. If $T=(p-1)/m$, then there are $m(p-1)$ free orbits, and $2m$ ideal orbits.

A3 If $(d/p)=0$, then $(p)=P_1^2$. The periods of orbits are computed as follows. Let $\lambda=(a+b\sqrt{D})/2$ (with a and b both even if $D\not\equiv 1 \pmod 4$). We have two cases (when p=2 the first case always holds):

1 - If $a\equiv 2 \pmod p$, there are $p-1$ ideal orbits of period 1, and $p-1$ free orbits of period p (or $p(p-1)$ of period 1). Each free orbit belongs to the same residue class modulo P_1.

2 - If $a\equiv -2 \pmod p$, there are $(p-1)/2$ ideal orbits of period 2, and $(p-1)/2$ free orbits of period $2p$ (or $p(p-1)/2$ of period 2). Each free orbit belongs to two residue classes mod P_1.

When $(d/p)\neq 0$, to compute the exact period T, one must test the congruence (2.4) (with $g=p$ and $(J,g)=(1)$) for all divisors of $p-(d/p)$. For $(d/p)=0$, only two congruences need to be tested, one for free orbits and one for ideal orbits.

The next step is to determine initial conditions for orbits on prime lattices, which is done by using norms. The norm of an integer z, taken modulo p, is invariant under the map. Indeed the latter corresponds to multiplication by λ, the norm of λ is 1, and the norm is multiplicative: $N(zw)=N(z)N(w)$, $z,w \in R$. Then we can associate to each orbit a rational integer between 0 and $p-1$, namely the least non-negative residue of the norm of any one of its points.

Let now G' be the multiplicative group of the ring $G=G(p)=R/pR$ (i.e. the group of reduced residue classes modulo (p), represented by quadratic integers relatively prime to p). Let L be the subgroup of G' generated by λ. The factor group G'/L is then the group of free orbits, and to find initial conditions we must find all its elements. The norm maps G'/L homomorphically into the multiplicative group of $\mathbf{Z}/p\mathbf{Z}$. The kernel of this homomorphism consists of all orbits having unit norm, which we call the subgroup U of unit orbits.

The procedure of determining initial conditions will now develop in four stages. First, for any given (non zero) value of the norm we find initial conditions for just one free orbit. Second, we find generators for U and construct all its elements. Then we construct all remaining free orbits as cosets of U in G . Finally, for each ideal divisor P of (p) (if any), we find one element $\pi \in P-(p)$, and construct all ideal orbits on P by multiplying the orbit through π by U. (Ideal orbits are in $G-G'$, and can still be multiplied by elements of G'.)

The first two steps deserve some comments. A representative point z of an orbit of norm a, $0<a<p$, is a solution to the congruence $N(z)\equiv a(\text{mod } p)$. This congruence can be shown

to be solvable (in z) for all a's if $(d/p)\neq 0$, whereas for $(d/p)=0$ it is solvable only for those a's which are quadratic residues modulo p (Hecke, 1923, §47). Accordingly, we shall regroup the set of a's into quadratic residues and non-residues modulo p. Expanding z using the $\mathbf{Z}$-basis $[1,\omega]$ of R, we see that $N(z)$ is a binary quadratic form (the principal form of discriminant d)

$$\begin{aligned} &N(z)=x^2-Dy^2, && D\not\equiv 1 \pmod 4; \\ &N(z)=x^2+xy+((1-D)/4)y^2, && D\equiv 1 \pmod 4. \end{aligned} \tag{3.2}$$

All z of the form $z=xz_0$, $z_0\in R$, $x=1,\ldots,(p-1)/2$, yield values of the norm which are distinct (whence they belong to distinct orbits) but have the same quadratic character. Indeed, from (3.2) we see that $(N(xz_0)/p)=(x^2N(z_0)/p)=(N(z_0)/p)$. It is therefore sufficient to find two integers z_0 for which $(N(z_0)/p)=1$ and -1, respectively (the latter case only if $(d/p)\neq 0$, see above). For quadratic residues an obvious choice is $z_0=1$, and it can be shown [PV] that at least one integer of the form $z_0=x+\omega$, $x=0,\ldots,p-1$ must have norm which is congruent to a non-residue modulo p. Such z_0 is to be found by trial and error, in general, although in some cases the solution can be determined beforehand from simple considerations. For instance, if the discriminant d is even and $(d/p)=-(-1/p)$, then the desired z_0 is ω, since $(N(\omega)/p)=(-D/p)=(d/p)(-1/p)=-1$.

As to the second step, let us consider the case $(d/p)\neq 0$ (if $(d/p)=0$, the matter is simple, see [PV]). From propositions A1 and A2 above, one finds that the order m of the group U is $(p-(d/p))/T$, T being the common period of all orbits. If $(d/p)=-1$, (p) is prime in R, whence G is a finite field and G' is cyclic. Then U is also cyclic, being a subgroup of G'. If $(d/p)=1$, U is non-cyclic, in general, and must be decomposed into the product of cyclic groups. Finding generators for U amounts to finding integers z of norm congruent to 1, and whose rth power (for some suitable r dividing m) belongs to the orbit through λ

(which is the identity of U). It can be shown [PV] that this problem is reduced to finding an integral root of an rth order polynomial over $\mathbf{Z}/p\mathbf{Z}$, plus solving a linear congruence.

The theory of periodic orbits for moduli (g) which are not primes, is naturally built up from that of prime moduli. One must first solve the case in which g is a primary integer: $g=p^n$, p a prime. In this case the structure of periods follows a regular pattern, as n increases. Indeed the period of the orbits increases linearly with the lattice size, beyond a certain power n_0, as detailed below.

A4 Let T be the period of free orbits to the prime modulus (p), and n_0 the smallest integer n for which $\lambda^T \not\equiv 1(\mathrm{mod}\ (p^n))$. Then, for odd primes p, the period T_n of all free orbits modulo (p^n) is the same: $T_n = Tp^{n-n_0+1}$. If $p=2$ the period is one of the numbers $T, Tp, Tp^2, \ldots, Tp^{n-n_0+1}$.

A5 Let T_n and n_0 be as above. Then orbits on any ideal factor of (p^n) which is not divisible by (p) have the same period $T_n' = T_n/a$ where $a=a(n)$ is either 1 or p. If p is an odd prime, then $a(n)=a(n_0+1)$, for all $n>n_0$.

In the general case of a composite modulus $(g)=p_1^{\alpha_1} \cdots p_t^{\alpha_t}$, one must first determine the solutions T_k to the congruences for primary moduli

$$\lambda^T z \equiv z(\mathrm{mod}\ (p_k^{\alpha_k})), \qquad k=1, \ldots, t. \tag{3.3}$$

Then the period of the orbit through z (for the modulus (g)), is just the least common multiple of the T_k's, from the Chinese remainder theorem. Since the solutions T_k of (3.3) depend only on $(z, p_k^{\alpha_k})$, it follows that all free orbits have the same period, as do ideal orbits belonging to conjugate ideal factors of the modulus (g).

4. PERIODIC ORBITS OF GIVEN PERIOD

To construct all orbits of a prescribed period n one must first determine the smallest integer M for which $G(M)$ contains all of them, and then use the techniques developed in §3 to locate them. $M=M(n)$ is called the *maximal modulus* (for the period n). Not all points of $G(M)$ need have period (dividing) n. Those which do turn out to form an ideal divisor of (M) (see below). To determine the maximal modulus we must solve the congruence

$$\lambda^N z \equiv z(\text{mod }(M)) \text{ or } v_n z \equiv 0(\text{mod }(M)), \qquad v_n=\lambda^N-1, \tag{4.1}$$

for the largest M for which at least one z exists such that (M,z) is not a rational integral ideal (except (1)). This guarantees that for no proper rational divisor g of M can $G(g)$ contain all the orbits of period n. The item to be considered here is the norm of v_n (cf. (4.1)).

A6 Let $|N(v_n)|=mr^2$, with r square-free. Then the maximal modulus is $M=mr$. Moreover, if $g|m$, then all points of $G(g)$ have period dividing n, and vice-versa.

A7 All orbits whose period divides n form a principal ideal $O=O(n)$, which divides the corresponding maximal modulus $M(n)$. We have $O=(\sqrt{D})$ for even n, while when n is odd, $O=((\lambda-1)/m_1)$, where m_1 is the largest rational integer dividing $\lambda-1$. Then O is an ambiguous ideal, i.e. it is equal to its conjugate and has no rational divisors.

A8 Let $\lambda>0$. Then $O=(1)$ if and only if the period is odd and $\lambda=\eta^2$, where η is a unit of norm -1.

Note that in order to have $O=(1)$ (for $\lambda>0$), the fundamental unit of R must have norm -1 (the fundamental unit is the generator of the group of all units, apart from sign considerations). This question can be decided case by case, using continued fractions (Wright, 1939, p.36). The problem of determining the sign of the norm of the fundamental unit directly from the discriminant d of R is more difficult, and still not completely solved. It is known that the norm of the fundamental unit is negative if d has only one prime divisor, and positive if d is divisible by primes of the form $4n-1$. Some other partial results are also available (see Cohn, 1978, p.105).

The maximal modulus M can be computed by means of a recursive formula, and it turns out that M is an exponentially increasing function of the period n [PV]. This fact typifies the numerical difficulties one encounters when attempting to construct unstable orbits.

5. CONCLUDING REMARKS

The above analysis allows one to classify and construct all periodic orbits of any given Anosov diffeomorphism, once the conjugacy function to the corresponding toral automorphisms is known. Although most periodic orbits are found to be structureless, orbits with special properties exist, which can be recognized by their number-theoretical properties and used for computational purposes.

Examples are ideal orbits of ''split'' prime lattices, which are distributed uniformly on the torus (when their period is maximal), since they are confined to invariant sublattices. On the finite set of ramified prime lattices, ''regular'' orbits are to be found, such as ideal fixed points and highly correlated free orbits (cf. A3 above). The maximal moduli M support ordered orbits, i.e. orbits conjugate to rotations, which are periodic approximants to invariant Cantor sets (Percival, 1978). The points of these orbits are clustered about periodic orbits of much smaller period, which accounts for the large size of the lattices which accomodate them.

The arithmetical methods we have described do not provide direct information about asymptotic properties of periodic orbits (such as the rate at which the number of orbits grows with the period). In this respect it is worth mentioning that analogous problems have been considered for the case of Anosov flows (geodesic flows on manifolds of everywhere negative curvature, see Parry and Pollicott, 1983, and references therein), employing concepts and techniques borrowed from analytic number theory. Similar methods are likely to be applicable also to the case of Anosov diffeomorphisms.

The generalization of the above techniques to the case of periodic orbits of Anosov diffeomorphisms of the n-dimensional torus is conceptually simple, but it will unavoidably involve greater computational difficulties. The roots of the characteristic equations of matrices of $SL_n(\mathbf{Z})$ are now algebraic integers θ of degree n over $\mathbf{Q}$. The relevant ring is $\mathbf{Z}[\theta]$ (see Taussky, 1978). Using the n-dimensional version of Eq. (2.2), one can again construct matrix classes starting from ideal classes in $\mathbf{Z}[\theta]$, and then consider the factorization of rational ideals in $\mathbf{Z}[\theta]$. The factorization algorithms are, of course, more complicated than those for quadratic fields, but they can be implemented on a computer.

REFERENCES

Cohn, H., 1962, "A Second Course in Number Theory", John Wiley & Sons, New York. [Reprinted by Dover, New York (1980) with the title "Advanced Number Theory".]

Cohn, H., 1978, "A Classical Invitation to Algebraic Numbers and Class Fields", Springer Verlag, New York.

Franks, J., 1970, "Anosov Diffeomorphisms", Proc. of Symp. in Pure Math., Vol. XIV, American Mathematical Society, Providence, R.I., p. 61.

Latimer, G. C., and MacDuffee, C. C., 1933, Ann. Math. **34** 316.

Manning, A., 1975, "Classification of Anosov Diffeomorphisms on Tori", Lecture Notes in Mathematics No. 468, Springer-Verlag, New York 26.

Hecke, E., 1923, "Vorlesung Uber die Theorie der Algebraischen Zahlen", Akademische

Verlagsgesellschaft, Leipzig. English translation: "Lectures on the Theory of Algebraic Numbers", Springer-Verlag, New York (1981).

Parry, W. and Pollicott, M., 1983, Ann. of Math. **118** 573.

Percival, I. C., 1979, "Variational Principles for Invariant Tori and Cantori", American Institute of Physics Conf. Proc. **57**, A.I.P., New York, p. 302.

Percival, I. C. and Vivaldi, F., 1987, "Arithmetical Properties of Strongly Chaotic Motions", preprint QMC DYN 86-2, Queen Mary College, London (1986). To be published in Physica D (1987).

Taussky, O., 1951, Pacific Journal of Mathematics **1** 127.

Taussky, O., 1978, "Introduction into Connections Between Algebraic Number Theory and Integral Matrices", appendix to Cohn, 1978 (see above).

Wright, H. N., 1939, "First Course in Theory of Numbers", John Wiley & Sons, New York, p. 36.

LIMITATIONS OF THE RABI MODEL FOR RYDBERG TRANSITIONS

P L KNIGHT AND S J D PHOENIX

1. INTRODUCTION

A two-level system driven by a classical single frequency field in rotating-wave approximation is a basic model for resonance in quantum optics (Allen and Eberly 1975). The probability of making a transition depends sinusoidally on time with a characteristic frequency called the Rabi frequency Ω

$$\Omega = d.E/\hbar \tag{1.1}$$

where d is the transition dipole moment and E is the amplitude of the external field. It is known from the work of Beloborov et al 1976 and Milonni et al 1983 that the semi-classical driven two-level system exhibits chaotic dynamics provided the rotating-wave approximation is not made. The two-level system driven by a quantized single mode field is also characterized by a Rabi frequency and its dynamics can be solved exactly in rotating-wave approximation. The quantized model however exhibits quantum mechanical dephasing of Rabi oscillations, revivals and an apparent disordered and irregular evolution at long times when photon statistics are properly described (see e.g. Eberly et al 1980, Narozhny et al 1981 and Barnett et al 1986). We shall be concerned in this article with these purely quantum effects and how they modify the semiclassical Rabi model. The relationship between the fully quantized model (known as the Jaynes-Cummings model) and the chaotic semiclassical model is not properly understood. Graham and Höhnerbach (1984) have made extensive studies of this problem and have gone beyond the rotating wave approximation to study the collapses and revivals. The experimental observation of quantum collapses and revivals has recently been reported by Walther and coworkers (1987) using highly excited Rydberg atoms in a superconducting high-Q cavity. In this article, we examine the photon statistics of the Jaynes-Cummings model and the use of entropy concepts to describe the apparently disordered collapses.

2. DESCRIPTION OF JAYNES-CUMMINGS MODEL

The Jaynes-Cummings model consists of a two-level atomic system driven by a single quantized electromagnetic field mode. The Hamiltonian describing the entire system is

$$H=H_a+H_f+V \tag{2.1}$$

where H_a is the atomic two-level Hamiltonian $H_a=\frac{1}{2}\hbar\omega_0\sigma_3$, H_f

is the field Hamiltonian $H_f = \hbar\omega a^\dagger a$ and V the coupling in rotating-wave approximation

$$V = \hbar\lambda(\sigma_- a^\dagger + \sigma_+ a) \tag{2.2}$$

where the inversion operator $\sigma_3 = |e\rangle\langle e| - |g\rangle\langle g|$, the lowering operator $\sigma_- = |g\rangle\langle e|$ and the raising operator $\sigma_+ = |e\rangle\langle g|$ are described in terms of the excited $|e\rangle$ and ground state $|g\rangle$ eigenvectors. The field frequency ω will be set equal to the atomic transition frequency ω_0, and the field annihilation and creation operators are a and $a^\dagger$. Finally, the atom-field coupling constant is λ. We use a density matrix method to describe the influence of photon statistics in the JCM (Stenholm 1973). The time-evolution operator U(t) derived from eq.(2.1) enables us to write the atom-field density matrix as

$$\rho(t) = U(t)\rho(0)U^\dagger(t). \tag{2.3}$$

In this article we will concentrate on the field properties and particularly on the reduced density operator for the field $\rho_f(t) = \mathrm{Tr}_a\{\rho(t)\}$. We have

$$\begin{aligned} U^\dagger(t) &= \exp(+iVt/\hbar) = \exp i\lambda t(a^\dagger\sigma_- + a\sigma_+) \\ &= \sum_{n=0}^{\infty} \frac{(i\lambda t)^n}{n!} \begin{pmatrix} 0 & a \\ a^\dagger & 0 \end{pmatrix}^n \end{aligned} \tag{2.4}$$

in the atomic basis $|e\rangle, |g\rangle$. Then using

$$\begin{pmatrix} 0 & a \\ a^\dagger & 0 \end{pmatrix}^{2m} = \begin{pmatrix} (aa^\dagger)^m & 0 \\ 0 & (a^\dagger a)^m \end{pmatrix}$$

$$\begin{pmatrix} 0 & a \\ a^\dagger & 0 \end{pmatrix}^{2m+1} = \begin{pmatrix} 0 & (aa^\dagger)^m a \\ (a^\dagger a)^m a^\dagger & 0 \end{pmatrix}$$

we find

$$\mathcal{U}^\dagger(t) = \sum_{m=0}^{\infty} \frac{(i\lambda t)^{2m}}{(2m)!} \begin{pmatrix} (aa^\dagger)^m & 0 \\ 0 & (a^\dagger a)^m \end{pmatrix} + \sum_{m=1}^{\infty} \frac{(i\lambda t)^{2m-1}}{(2m-1)!} \begin{pmatrix} 0 & (aa^\dagger)^{m-1} a \\ (a^\dagger a)^{m-1} a^\dagger & 0 \end{pmatrix}$$

so that

$$\mathcal{U}^\dagger(t) = \begin{pmatrix} \cos[\lambda t (aa^\dagger)^{1/2}] & \dfrac{i \sin[\lambda t (aa^\dagger)^{1/2}]\, a}{(aa^\dagger)^{1/2}} \\ i \dfrac{\sin[\lambda t (a^\dagger a)^{1/2}]\, a^\dagger}{(a^\dagger a)^{1/2}} & \cos[\lambda t (a^\dagger a)^{1/2}] \end{pmatrix} \tag{2.5}$$

The reduced density operator for the field is given by

$$\rho_f(t) = \mathrm{Tr}_a\{U(t)\rho(0)U^\dagger(t)\}. \tag{2.6}$$

If at time t=0 the atom and field are uncorrelated, with the atom initially in the excited state $|e\rangle$, we find

$$\rho_f(t) = \mathrm{Tr}_a\left\{U(t)\begin{pmatrix}\rho_f(0) & 0\\ 0 & 0\end{pmatrix}U^{+}(t)\right\}$$

$$= C\rho_f(0)C^{+} + S\rho_f(0)S^{+} \qquad (2.7)$$

where

$$c=\cos[\lambda t(aa^{+})^{1/2}] \qquad (2.8a)$$

$$s=a^{+}\frac{\sin[\lambda t(aa^{+})^{1/2}]}{(aa^{+})^{1/2}}. \qquad (2.8b)$$

In the rotating-wave approximation, the Jaynes-Cummings Hamiltonian conserves excitation number N

$$N=a^{+}a+\tfrac{1}{2}(1+\sigma_3) \qquad (2.9)$$

so that atomic and field excitations are related by

$$\langle a^{+}a\rangle=\langle N\rangle-\tfrac{1}{2}(1+\langle\sigma_3\rangle) \qquad (2.10)$$

3. FIELD EVOLUTION

If the single-mode radiation field were prepared initially in a number state so that $\rho_f(0)=|n\rangle\langle n|$, we find from eq. (2.7)

$$\rho_f(t)=C|n\rangle\langle n|C^{\dagger} + S|n\rangle\langle n|S^{\dagger} \tag{3.1}$$

$$=\cos^2(\Omega_{n+1}t/2)|n\rangle\langle n| + \sin^2(\Omega_{n+1}t/2)|n+1\rangle\langle n+1|$$

where Ω_{n+1} is the Rabi frequency

$$\Omega_{n+1}=2\lambda(n+1)^{1/2} \tag{3.2}$$

An excited atom interacting with a vacuum state single mode field (n=0) generates a reduced density matrix for the field

$$\rho_f(t)=\cos^2(\Omega_1 t/2)|0\rangle\langle 0|+\sin^2(\Omega_1 t/2)|1\rangle\langle 1| \tag{3.3}$$

exhibiting spontaneous one-photon Rabi oscillations. The average photon number oscillates at Ω_1 :

$$\langle a^{\dagger}a\rangle=\sin^2(\Omega_1 t/2) \tag{3.4}$$

If the field were prepared in a thermal state,

$$\rho_f(0)=\sum_{n=0}^{\infty} p(n)\ |n\rangle\langle n| \tag{3.5}$$

with

$$p(n)=\bar{n}^{n}/(1+\bar{n})^{1+n} \tag{3.6}$$

then the reduced density operator for the field at time t is

$$\begin{aligned}\rho_f(t)&=\sum_{n=0}^{\infty} p(n)\{c\,|n\rangle\langle n|\,c^{+}+s\,|n\rangle\langle n|\,s^{+}\}\\ &=p(0)\ \cos^2(\Omega_1 t/2)\,|0\rangle\langle 0|\\ &+\sum_{n=1}^{\infty}\Big\{p(n)\cos^2(\Omega_{n+1}t/2)\\ &\qquad+p(n-1)\sin^2(\Omega_n t/2)\Big\}|n\rangle\langle n|.\end{aligned} \tag{3.7}$$

We write eq.(3.7) in the abbreviated form

$$\rho_f(t)=\sum_{n=0}^{\infty}\Pi(n,t)\ |\ n\rangle\langle n| \tag{3.8}$$

which represents a mixed state with time-dependent coefficients . The average photon number is

$$\langle a^{+}a\rangle=\sum_{n=0}^{\infty} n\Pi(n,t). \tag{3.9}$$

If the field were prepared in a coherent state, the density

matrix possesses off-diagonal elements in the number state basis. Nevertheless, these do not contribute to the sum of diagonal elements of $\rho_f\, a^\dagger a$ which make up $\langle a^\dagger a\rangle$. We find for this case

$$\langle a^\dagger a\rangle=\sum_{n=0}^{\infty} n\{\langle n|\,(c|\alpha\rangle\langle\alpha|\,c^\dagger+S|\alpha\rangle\langle\alpha|\,S^\dagger)\;n\rangle\}$$

$$=\sum_{n=0}^{\infty} n\Pi(n,t) \tag{3.10}$$

as before, but with

$$\Pi(0,t)=p(0)\;\cos^2(\Omega_1 t/2) \tag{3.11a}$$

$$\Pi(n,t)=p(n)\;\cos^2(\Omega_{n+1} t/2)+p(n-1)\sin^2(\Omega_n t/2) \tag{3.11b}$$

and

$$p(n)=\frac{\bar{n}^n}{n!}e^{-\bar{n}}\;. \tag{3.12}$$

4. FIELD ENTROPY IN THE JAYNES-CUMMINGS MODEL

Entropy is often used to characterize statistical disorder (we set aside anthropomorphic problems involving the subjectivity of statistical entropies (Denbigh and Denbigh 1985)) through the relation

$$S=-\sum_i p_i\,\ln p_i \tag{4.1}$$

where p_i is the probability of the occurrence of state i.

The quantum generalization we consider defines entropy through

$$S = -\mathrm{Tr}\{\rho \ln \rho\} \tag{4.2}$$

where the trace is taken over a convenient basis set. In this generalization the entropy of a pure state is zero, and the existence of a non-zero S describes in a sense the additional uncertainties over and above those inherent quantum uncertainties which exist even for a pure state. Aravind and Hirschfelder (1984) have calculated the entropy of the two-level atom in the JCM. Here we discuss the entropy of the field and find a number of new features in which the entropy oscillates, collapses and revives. In terms of the number states, the field entropy is given by

$$S_f = -\sum_{n=0}^{\infty} \langle n|\rho_f \ln \rho_f|n\rangle$$

$$= -\sum_{n=0}^{\infty} (\rho_f)_{nn} \ln(\rho_f)_{nn} \tag{4.3}$$

if the states $|n\rangle$ form the eigenket basis of ρ_f. In general this will not be so and the reduced density operator will have to be diagonalized. If this cannot be done, the number state basis result eq.(4.3) can still be used to define a Shannon information entropy (Shannon and Weaver 1963) which may be used for example to describe the photon number probability distributions. The field entropy is most easily calculated for the field initially in the vacuum state. From eq.(3.3)

$$S_f = -\{\cos^2(\Omega_1 t/2)\ln[\cos^2(\Omega_1 t/2)] + \sin^2(\Omega_1 t/2)\ln[\sin^2(\Omega_1 t/2)]. \quad (4.4)$$

The average photon number and the entropy are plotted in Fig.(1) for this simple case

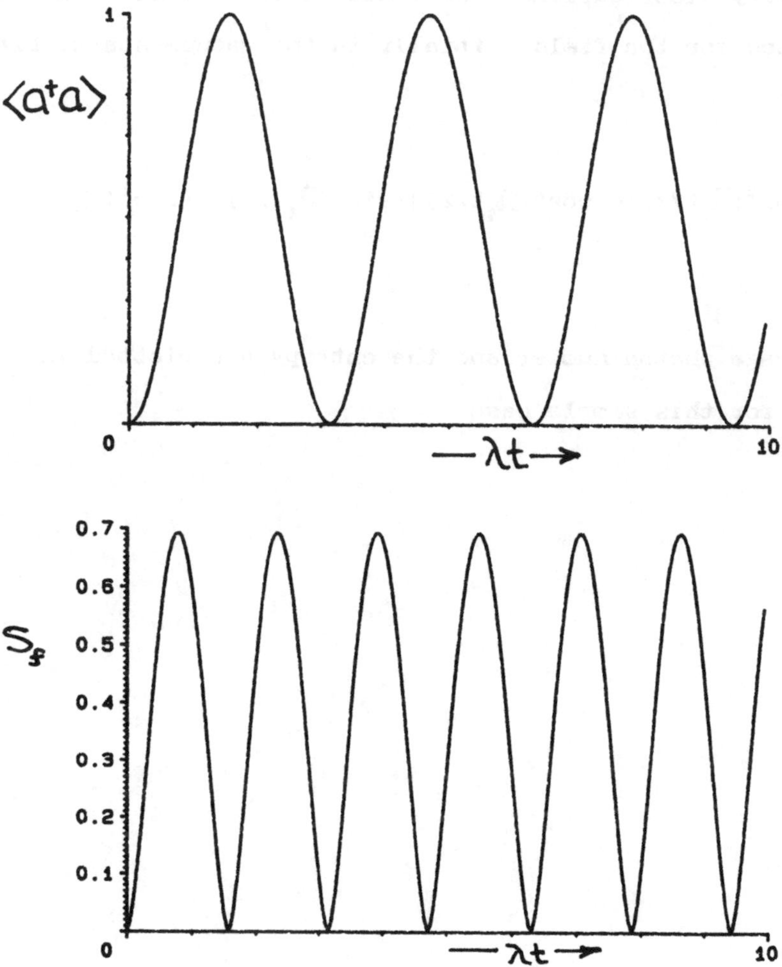

Fig. 1: Photon number and field entropy for vacuum Jaynes-Cummings time evolution.

The entropy oscillates at twice the frequency of the average photon number, but this is as expected because when the field is in either state $|0\rangle$ or $|1\rangle$ the entropy is zero. The regular evolution of the entropy underscores the reversibility of this simple model.

The thermal initial field is also diagonal in the number states and similarly leads to a simple expression for the field entropy

$$S = -\sum_{n=0}^{\infty} \Pi(n,t) \ln \Pi(n,t) \tag{4.5}$$

where the time-dependent factors $\Pi(n,t)$ are given by eq.(3.11) and

$$p(n) = \bar{n}^{n}/(1+\bar{n})^{1+n} \tag{4.6}$$

The mean photon number and the field entropy evolution for a thermal initial field state and an initially excited atom is plotted in figure 2 for $\bar{n}=15$. As noted before (Knight and Radmore 1982), the JCM evolution for this problem becomes more and more irregular as time evolves. The field entropy, on the other hand appears to exhibit some kind of slow modulation. It is also notable that the field entropy initially decreases when the atom begins to interact with the thermal field, indicating an initial ordering of the system.

The field entropy in the case of an atom interacting with an initially coherent field is expected to reflect the collapse and revival of Rabi oscillations. To avoid the problem of diagonalizing the density matrix, we employ the Shannon information definition of entropy in the number state basis (corresponding in this sense to a phase-diffused coherent light),

$$S = -\sum_{n=0}^{\infty} \Pi(n,t) \ln \Pi(n,t) \tag{4.7}$$

with the time-dependent factors given again by eq.(3.11) and the photon distribution p(n) given by the Poisson distribution. This is shown in fig.(3)

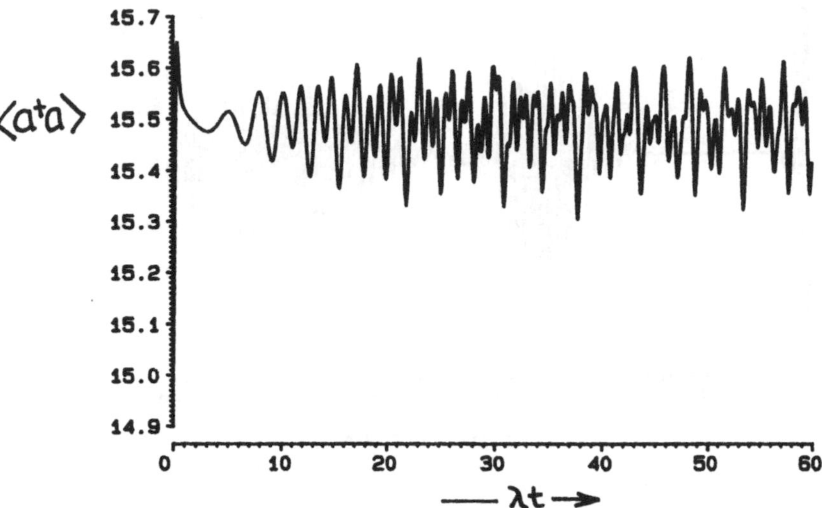

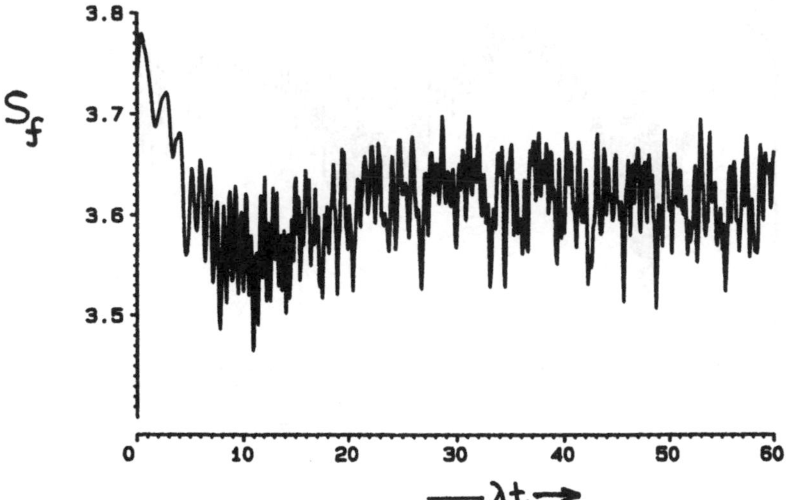

Fig.2: The evolution of (a) the average photon number and (b) the field entropy for an initially excited atom driven by a thermal field with initial average photon number $\bar{n}=15$.

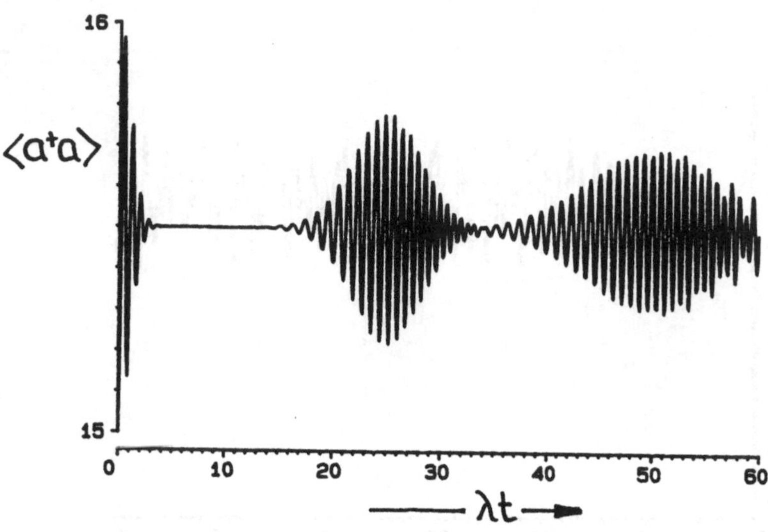

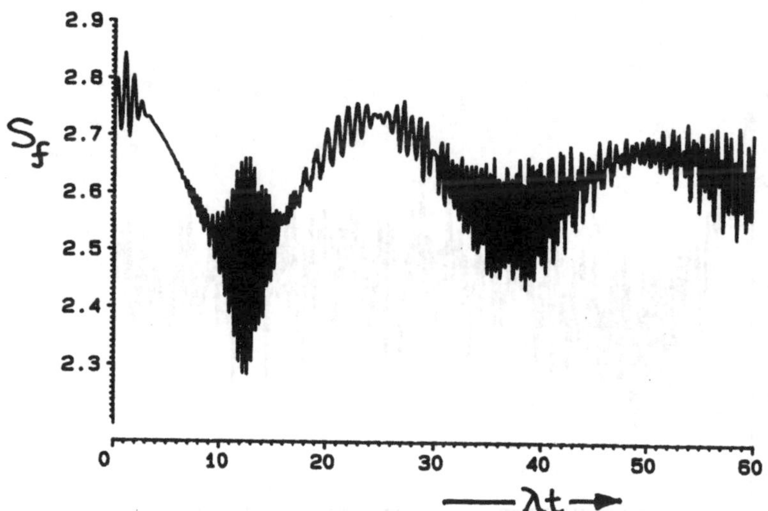

Fig. 3: The evolution of (a) the average photon number and (b) the field entropy for an initially excited atom interacting with an initially coherent (phase-diffused) field with average photon number $\bar{n}=15$.

We note from Fig.3 that the entropy decreases initially as for the case of thermal excitation and shows a strong periodic modulation at the revival period. Towards the end of the first collapse, detailed structure appears in S . The first maximum of the entropy envelope appears at the position of the first revival (here at $\lambda t \sim 25$) where the field most closely returns to its initial state. The decrease in the field entropy initially seems connected with the change in the photon statistics during the JCM evolution. The photon number distribution (given by $\Pi(n,t)$) becomes more structured and can be narrower than the initial Poissonian width. The increase in field entropy reflects this greater certainty in the photon statistics.

5. PHOTON NUMBER VARIANCES

The variance in the photon number operator $n=a^{\dagger}a$ also reflects both collapses and revivals. We define the variance $(\Delta n)^2$ through

$$(\Delta n)^2=[\langle a^{\dagger} aa^{\dagger} a\rangle-\langle a^{\dagger} a\rangle^2] \tag{5.1}$$

As all of these expectation values are computed at equal times they are simple to deduce from the density matrix. We find

$$\langle a^{+}a\rangle = \bar{n} + (1 - \langle\sigma_3(t)\rangle)/2 \tag{5.2}$$
$$= \sum_{n=0}^{\infty} \langle n|(c\rho_f(0)c^{+} + s\rho_f(0)s^{+})a^{+}a)|n\rangle$$
$$= \sum_{n=1}^{\infty} n\Pi(n,t),$$

where $\Pi(n,t) = p(n)\cos^2(\Omega_{n+1}t/2) + p(n-1)\sin^2(\Omega_n t/2)$.

Similarly

$$\langle a^{+}a\, a^{+}a\rangle = \sum_{n=1}^{\infty} n^2\Pi(n,t).$$

However we know that $\langle\sigma_3\rangle = \sum_{n=0}^{\infty} p(n)\cos(\Omega_{n+1}t)$,

and it is straightforward to show that

$$\sum_{n=0}^{\infty} p(n)\sin^2(\Omega_{n+1}t/2) = \frac{1}{2}(1 - \langle\sigma_3\rangle)$$

then we find for the term $\langle a^{+}aa^{+}a\rangle$

$$\langle a^{+}aa^{+}a\rangle = \overline{n^2} + \bar{n} - \sum_{n=1}^{\infty} np(n)\cos(\Omega_{n+1}t) + \frac{1}{2}(1 - \langle\sigma_3\rangle)$$

and for the variance

$$(\Delta n)^2 = (\Delta n)^2_{t=0} + \frac{1}{4}(1 - \langle\sigma_3\rangle^2) + \bar{n}\langle\sigma_3\rangle - \sum_{n=1}^{\infty} np(n)\cos\Omega_{n+1}t \tag{5.3}$$

We note that $\langle\sigma_3^2\rangle = 1$ so that we find

$$(\Delta n)^2 = (\Delta n)^2_{t=0} + \frac{1}{4}(\Delta\sigma_3)^2 + \bar{n}\langle\sigma_3\rangle - \sum_{n=1}^{\infty} np(n)\cos(\Omega_{n+1}t) \tag{5.4}$$

This can be written in the equivalent form

$$(\Delta n)^2 = (\Delta n)^2_{t=0} + \frac{1}{4}(\Delta\sigma_3)^2 + (\bar{n}+1)\langle\sigma_3\rangle + \frac{1}{\Omega_1^2}\frac{d^2}{dt^2}\langle\sigma_3\rangle\,. \tag{5.5}$$

For an initially coherent field p(n) is a Poisson distribution for which $\bar{n}p(n) = (n+1)p(n+1)$ so that

$$(\Delta n)^2 = \bar{n} + \frac{1}{4}(\Delta\sigma_3)^2 - 2\sum_{n=0}^{\infty} p(n)\sin[(\Omega_{n+1}+\Omega_{n+2})t/2]\sin[(\Omega_{n+2}-\Omega_{n+1})t/2]$$

In figure 4 we show the time-evolution of the photon number variance for the coherent state JCM for $\bar{n}=15$. Note the initial variance is 15 as expected for a Poisson field and increases and decreases with the Rabi oscillations with a significant sub-Poissonian minimum value. This last feature is again purely quantum mechanical and indicates transient sub-Poissonian photon statistics could be generated in the Rydberg atom maser.

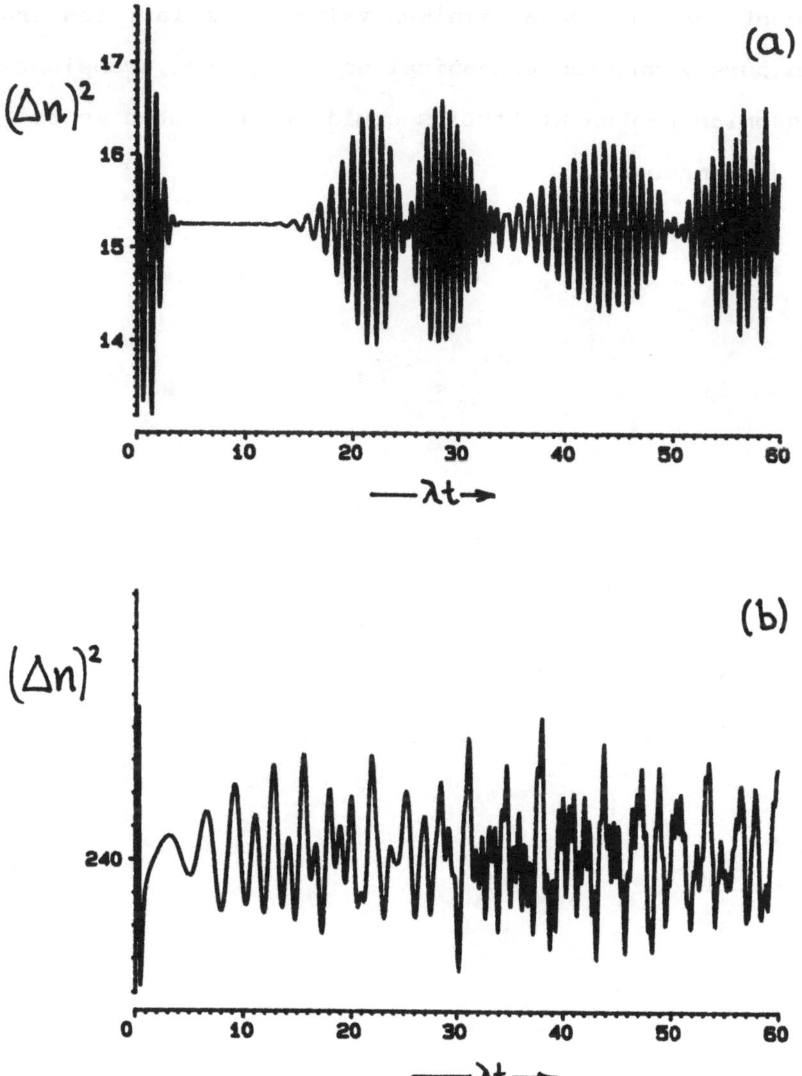

Figure 4-Time evolution of photon variance for (a) coherent state JCM and (b) thermal state JCM, each with an initial mean photon number of 15.

Also worth noting in figure 4 is that the quasi-steady state region after the first collapse has a significantly increased variance. We note from (5.5) that during the collapse when $\langle\sigma_3\rangle$ is approximately zero that

$$(\Delta n)^2 \approx (\Delta n)^2_{t=0} + \frac{1}{4} \tag{5.6}$$

and this is confirmed by the numerical plots. Further, the revivals in the variances are modulated by an envelope which supresses the Rabi revivals in the variance at the point of maximum inversion revival. When the inversion is at a point of maximum revival the system returns most closely to its initial state which for fig.(4) has a photon number variance of 15.

The thermal field JCM shows enhanced photon variances as expected for an initially Bose-Einstein field with large fluctuations. Nevertheless the quantum revivals again modify the initial variances and this is shown in figure 4(b).

Given the current interest in squeezed field fluctuations in Rydberg atom masers (e.g. Knight 1986) it seems possible that experimentalists may be able to explore the photon statistics of the Jaynes-Cummings model. Provided that the cavity losses were negligible on the timescale of the quantum collapses (as in the Munich experiments of Walther et al) it seems that some of the effects shown here could be important.

This work was supported in part by the Science and Engineering Research Council. We are grateful to S.M.Barnett for discussions.

REFERENCES

Allen L. and Eberly J.H. (1975),"Optical Resonance and Two-level Atoms" Wiley, New York.

Aravind P.K. and Hirschfelder J.O. (1984), J.Phys.Chem.88, 4788.

Barnett S.M., Filipowicz P., Javanainen J., Knight P.L. and Meystre P. (1986) in "Frontiers in Quantum Optics" eds. E.R. Pike and S. Sarkar,Adam Hilger, Bristol.

Beloborov P.I., Zaslavski G.M. and Tartakovski G.K. (1976),Zh.Eksp.Teor.Fiz.71,1799

Denbigh K.G. and Denbigh J.S. (1985) "Entropy in Relation to Incomplete Knowledge", Cambridge University Press, Cambridge.

Eberly J.H. Narozhny N.B. Sanchez-Mondragon J.J. (1980) Phys.Rev.Lett.44,1323.

Graham R. and Höhnerbach M. (1984), Z.Phys.B.57,233.

Knight P.L. and Radmore P.M. (1982),Phys.Rev.A26,676

Knight P.L. (1986),Physica Scripta T12,51

Milonni P.W, Ackerhalt J.R. and Galbraith H.W. (1983),Phys.Rev.Lett.50,966

Narozhny N.B, Sanchez-Mondragon J.J. and Eberly J.H. (1981),Phys.Rev.A23,236

Shannon C.E. and Weaver W. (1963),"The Mathematical Theory of Communication",Illini books,Urbana

Stenholm S. (1973) Phys.Rep.C6,1.

Walther H. (1987) Private Communication.

QUASI-PROBABILITY DISTRIBUTIONS IN ASTABLE DISSIPATIVE QUANTUM SYSTEMS

J S SATCHELL, SARBEN SARKAR AND H J CARMICHAEL

1. INTRODUCTION: MOTIVATION FOR THE USE OF QUASI-PROBABILITY METHODS

Instabilities and chaos in Hamiltonian and dissipative quantum systems have been the subject of much study (Casati, 1983; Sarkar, 1986a; Pike and Sarkar, 1986b). There has been some divergence of the quantities of interest in the two types of systems. Probabilistic ideas, however, arise naturally in all quantum mechanical systems: the wavefunction is blurred over volumes of order $\hbar^N$ in the appropriate 2N dimensional phase phase space for the system where $\hbar$ is Planck's constant. We can no longer speak of variables of the system having well defined values. Instead we must deal with variables which are operators and evaluate their expectation values. When the equivalent classical system has instabilities (bifurcations or chaos) the indeterminacy of the variables in the corresponding quantum system is expected to radically alter the behaviour on small enough scales.

The study of Wigner functions in Hamiltonian systems (Berry, 1983) has given some insight into these questions. Dissipative systems still are comparatively less well understood. The theoretical techniques that are required for dissipative systems are different in flavour. In this article we will introduce in a gentle way the pot-pourri of methods that have been used in the study of these problems. We will cast a critical eye over the application of these techniques to the study of instabilities in dissipative systems.

The state of a quantum dissipative system is defined by its density operator. The time development of this operator is given by a quantum mechanical master equation (ME) (Haken, 1970; Gardiner, 1983). MEs are not usually directly soluble; some information on low order expectation values may be found by making semi-classical decorrelation approximations for large systems. An alternative method was developed in the 1960s by Glauber, Haken and others based on the use of the solutions of Fokker-Planck equations (FPE) for quasi-probability distributions (Carmichael 1986). These methods have, over the years, been applied extensively to models which arise in quantum optics.

In the next section we briefly show how a generalised FPE can be associated with a master equation. This step is exact, provided the appropriate boundary conditions on the quasi-probability hold, but the resulting equation is generally intractable. The generalised FPE is a multi-dimensional partial differential equation (PDE) giving the time development of a quasi-probability distribution as a function of the phase space variables.

Difficulties arise because derivatives of arbitrary orders may appear. To obtain a FPE, which only involves derivatives of up to second order, it is necessary to adopt a truncation approximation. If the higher derivatives fall off as the inverse of some large parameter (usually the system size) we can truncate the expansion at second order. This is a somewhat uncontrolled approximation, and will be discussed below.

Even when we have a FPE there are still a number of difficulties. The FPE is a multidimensional PDE, and, as such, frequently resists attempts at direct analytic or numerical solution. When the system has stable fixed points we can examine the solution in their vicinity by linearised analysis: this obviously breaks down when the fixed points are unstable. Further difficulties arise because the quasi-probabilities sometimes become negative or highly singular. In these cases progress is generally difficult.

When the distributions are described by well behaved positive functions (which requires positive definite diffusion in the FPE) we can use the powerful techniques of stochastic differential equations (Gardiner 1983). It can be shown that there is a one to one correspondence between an FPE with positive definite diffusion and a set of stochastic differential equations (SDEs). Moments of the solution of the FPE can be obtained from moments of an ensemble of trajectories obeying the SDEs. Although the SDEs are not usually directly soluble, they can be numerically simulated, and estimates obtained for operator expectation values. In chaotic systems one can study the fractal and scaling properties of the steady state probability distributions, and also calculate Lyapunov exponents.

For these reasons there has been a strong motivation to find new representations of quasi-distributions with good computational properties. One notable example is due to Drummond and Gardiner (1980). They have shown by construction for bosonic systems that there is one distribution which always has positive definite diffusion (if a generalised FPE having no more than second order derivatives exists). This is the so-called positive-P distribution. It requires the use of an enlarged phase space, but the additional degrees of freedom have zero mean value. Until recently the main use of this distribution has been to provide formal justification for the use of linearisation about fixed points when the related Glauber P distribution is singular. Because the positive-P distribution always gives positive definite diffusion in the FPE, it might be hoped that this would allow us to treat highly nonlinear systems also. Unfortunately there are practical difficulties relating to instabilities in the additional degrees of freedom that have been introduced. It is not clear whether these are surmountable or not. It may very well be that in its most interesting cases the technique breaks down.

We will consider in detail two related models, drawn from the area of quantum optics to illustrate the above points. They are the single mode laser and absorptive optical bistability.

The single mode laser shows chaos for large pump parameters (Haken 1975). We have used SDEs to study the effects of intrinsic noise on the chaos (Sarkar, Satchell and Carmichael, 1986b) by examining how it changes the static properties (e.g. fractal dimension and asymptotic probability distribution) and dynamic properties (e.g. Lyapunov exponent) (Sarkar and

Satchell, 1987a). As a bonus we can also shed some light on the interesting question of algorithmic complexity in quantum chaos (Ford, 1986).

Most of these calculations have been performed with the Wigner distribution. This has 'effective positive definite diffusion', but potential problems exist for other systems. We use the term 'effective positive definite diffusion' when the region of non-positive diffusion is very rarely (if ever) visited by trajectories. We have also examined the positive-P distribution and discovered that, even at the deterministic level, the equations are liable to show pathological behaviour.

In absorptive optical bistability there can be two possible values of output intensity for a given input intensity. Hysteresis and two co-existing fixed points when C, the co-operativity parameter (to be defined later) is greater than 4. When C = 4 there is a critical point of neutral stability and enhanced response to fluctuations. It is possible to numerically solve the master equation for the bad cavity case for small numbers of atoms, and we compare the results of the FPE calculations with these. The FPE works well for cases with stable fixed points, but difficulties with some of the quasi-probability distributions became apparent for the critical case. Again the pathological behaviour of the positive-P solution is found.

2. GENERALISED FOKKER-PLANCK EQUATIONS AND QUASI-PROBABILITY DISTRIBUTIONS

The master equation can be related to a FPE for a quasi-probability distribution by the use of a characteristic function. The characteristic function has the form of a generating function for the quasi-probability distribution. Different choices of operator orderings will give rise to different characteristic functions. The characteristic function is then related to the quasi-probability distribution by an integral transform. Many possible distributions can be obtained in this way. However we will study two here. The Wigner distribution is obtained from the characteristic function for symmetrically ordered operators, and the positive-P distribution from the characteristic function for normally ordered operators.

The master equations for quantum optics, that we will study in the examples below, contain two types of operators. The field operators a and a^+ are associated with a single mode of the radiation field, and obey the commutation relation

$$[a,a^+] = 1 \tag{1}$$

This field interacts with a collection of two level atoms. If we choose a basis for the states of the μth atom in which the excited and ground states are represented by

$$\begin{pmatrix}1\\0\end{pmatrix}, \begin{pmatrix}0\\1\end{pmatrix}$$

the atomic operators σ_μ^+, σ_μ^- and σ_μ^3 for the μth atom have the form of the Pauli spin matrices

$$\sigma_\mu^+ = \begin{pmatrix} 0 & 1 \\ 0 & 0 \end{pmatrix} , \quad \sigma_\mu^- = \begin{pmatrix} 0 & 0 \\ 1 & 0 \end{pmatrix} , \quad \sigma_\mu^3 = \begin{pmatrix} 1 & 0 \\ 0 & -1 \end{pmatrix} \tag{2}$$

To obtain the Wigner distribution we introduce the operators

$$\chi_\mu(\xi,\xi^*,\eta) = \exp i[\xi^*\sigma_\mu^+ \exp(ik.x_\mu) + \eta\sigma_\mu^3 + \xi\sigma_\mu^- \exp(-ik.x_\mu)] \tag{3a}$$

$$\bar{\chi} = \exp i(\zeta^* a^+ + \zeta a) \tag{3b}$$

$$\chi = \bar{\chi} \prod_{\mu=1}^{N} \chi_\mu \tag{3c}$$

where ρ is the density matrix. The characteristic function is

$$C_N(\xi,\xi^*,\eta,\zeta,\zeta^*) = \mathrm{Tr}(\chi) \tag{4}$$

The generalised Wigner distribution P is then defined by

$$C_N(\xi,\xi^*,\eta,\zeta,\zeta^*) = \int \ldots \int d\bar{\xi}\ d\bar{\xi}^*\ d\bar{\eta}\ d\bar{\zeta}\ d\bar{\zeta}^*\ P(\bar{\xi},\bar{\xi}^*,\bar{\eta},\bar{\zeta},\bar{\zeta}^*) \times \exp i(\bar{\zeta}\zeta + \bar{\zeta}^*\zeta^* + \bar{\xi}^*\xi + \bar{\xi}\xi + \bar{\eta}\eta) \tag{5}$$

and

$$\zeta = \zeta_x + i\zeta_y \quad , \quad \xi = \xi_x + i\xi_y$$

Operator expectation values can be obtained by taking corresponding moments of the distribution P. Unfortunately the Wigner distribution is not necessarily positive, and so an FPE obtained for it does not always have positive definite diffusion.

The derivation of the positive-P distribution proceeds similarly. If $S = \{0_i, 0_i^+\}$ is a complete set of operators for a system (in the sense that any other operators can be expressed as a polynomial in these operators) then the characteristic function associated with S is

$$\chi_S = \mathrm{Tr}\left(\rho \prod_{j=1}^{n} \exp(i\lambda_j' 0_j^+) \prod_{j'=1}^{n} \exp(i\lambda_{j'} 0_{j'})\right) \tag{6}$$

(For Hermitian operators $0_j^+ = 0_j$ and $\lambda_j' = 0$.)

A class of distribution functions $P_S(\{\alpha_j,\alpha_j'\})$ can be related to this by

$$\chi_S = \int \prod_{i=1}^{n} d\mu(\alpha_i,\alpha_i') \exp\left(i \sum_{j=1}^{n} \lambda_j\alpha_j\right) \exp\left(i \sum_{k=1}^{n} \lambda_k'\alpha_k'\right) P_s(\{\alpha_i,\alpha_i'\}) \tag{7}$$

where μ is the integration measure. The well known Glauber P distribution corresponds to

$$d\mu\ (\alpha_i,\alpha_i') = d^2\alpha_i\ \delta_{\alpha_i^*\alpha_i'} = d(\mathrm{Re}\alpha_i)\ d(\mathrm{Im}\alpha_i)\ \delta_{\alpha_i^*\alpha_i'} \tag{8}$$

For this distribution the support is confined to $\alpha_i^* = \alpha_i'$, i.e. the pairs of variables are not independent and Eqn (7) becomes a Fourier transform. If we take

$$d\mu(\alpha_i,\alpha_i') = d^2\alpha_i \; d^2\alpha_i' \quad , \tag{9}$$

the positive-P distribution is obtained. Normally ordered operator expectation values can be found from suitable moments of the positive P distribution. There is no explicit transform which expresses P in terms of the characteristic function. In fact, the existence or uniqueness of such a P cannot be proved in general. For the case where $S = \{a,a^+\}$, and a and a^+ are the annihilation and creation operators for the harmonic oscillator, it can be shown that there exists a suitable P_S. A possible $P_S(\alpha,\alpha')$ has the form

$$P_S(\alpha,\alpha') = \frac{1}{4\pi^2} \exp(-\tfrac{1}{4}|\alpha'-\alpha^*|^2) \langle \tfrac{1}{2}(\alpha'+\alpha^*)|\rho| \tfrac{1}{2}(\alpha'+\alpha^*)\rangle \tag{10}$$

where the coherent state $|\beta\rangle$ is such that

$$\alpha|\beta\rangle = \beta|\beta\rangle \quad .$$

However when S contains atomic operators no such simple form for PS is known. The master equation for ρ implies in general a partial differential equation with derivatives of all orders for PS. In these cases further progress is only possible if we make a truncation approximation, as discussed in the next section. If a truncation approximation is made the equation for PS reduces to a second order one in $\{li,li'\}$, a FPE.

3. SYSTEM SIZE EXPANSIONS AND FPEs

What we call the generalised FPE is known as the Kramers-Moyal expansion in the theory of classical stochastic processes. In principle it is possible to make a systematic expansion in some suitable parameter which scales as an inverse power of the size of the system under consideration (van Kampen, 1981) so that we obtain a description of fluctuations about some mean motion of the system. Our procedure for obtaining a nonlinear FPE is a straight truncation of the Kramers-Moyal expansion. This is not systematic, because even when noise terms are of order $1/\Omega$ where Ω is the system size, and terms higher in the order of derivatives have coefficients of order $1/\Omega^n$ (after transformation to intensive variables), the width of the distribution is still determined by $1/\Omega$. Consequently further Ω-dependence lies hidden in the derivatives (which contribute terms whose size depends on the sharpness of the distribution). However a system-size expansion is not easy to do for multi-variable generalised FPEs. (Occasionally, especially for purely bosonic systems, the FPE is exact and no truncation is needed. The anharmonic oscillator with analytic nonlinearity symmetric in the co-ordinate and conjugate momentum is a case in point (Satchell and Sarkar, 1986).)

If the system has stable fixed points, one can see that the contribution from the truncated terms falls off as $1/N^2$. We will illustrate this below for the case of absorptive optical bistability. In this case we have an explicit numerical solution for the master equation. When the fixed points are of neutral stability the drift (first derivative) term exactly vanishes at the fixed point. The truncated terms must now be compared with the diffusion (which goes as $1/N$) rather than the drift (independent of N). It is obvious that in this case the truncation approximation is more problematic. This too can be seen from our example. In fact a higher order Taylor expansion of the drift is needed and the power of 1/N characterising the noise for a systematic expansion changes.

The question of approximations is more difficult in the case of chaotic systems. Because of sensitive dependence on initial conditions (Lichtenberg and Lieberman, 1983), it is apparent that even good approximations soon break down in the sense that trajectories exponentially diverge from the 'true' ones. However we can invoke the 'shadowing theorem' to say that the trajectories remain near some true trajectory at all times. The system size expansion does not involve approximations to the drift (the deterministic motion), so the motion of a delta function initial probability distribution in the small noise limit will exactly follow the equivalent deterministic trajectory. However the use of a distribution is in some sense an approximation. The development of an initial distribution to cover the attractor will not be exact, nor indeed will the detailed form of the eventual steady state distribution. The truncated terms involve higher derivatives of the distribution function with respect to the variables. If the distribution being approximated is reasonably smooth then the approximation is at least self consistent. We can see from this that the system size expansion will break down if the support of the distribution becomes fractal. Fortunately, as we will see below the distributions show some evidence for becoming smooth at small enough scales.

In practice the form of the generalised FPE is sometimes inconveniently complex and one proceeds directly to the standard FPE. It is possible to use operator techniques in these cases and obtain the usual type of FPE directly. The derivation of the FPE for the Wigner distribution proceeds in this way (Sarkar, 1986a). It was first derived by Gronchi and Lugiato (1978) by somewhat different methods.

In the case of the positive-P generalised FPE we can show that derivatives of all orders occur with respect to the inversion variable. These are associated with difference rather than differential operations. The inversion variable only has support at integer values, corresponding to definite numbers of atoms being excited. We do not want such a fine grained description (nor are we able to solve the difference-partial differential equation that results). Instead we replace the delta functions in the inversion by a smooth envelope. In fact the retention of derivatives of up to second order is equivalent to fitting such a locally parabolic envelope. An important result of this approximation is that while the general FPE has to support outside the physical region, the approximate one does leak into 'nonphysical' regions of the inversion variable.

4. STOCHASTIC DIFFERENTIAL EQUATIONS AND QUASI-PROBABILITY DISTRIBUTIONS

In typical applications we have multidimensional PDEs, for which there are no acceptable numerical methods of solution. This limits us to the use of linearised analysis about fixed points or else the Ito decomposition (SDEs). These only exist if the diffusion is positive definite. Unfortunately, as already noted, this is not necessarily so for all quasi-probability distributions.

The Ito calculus allows us to make a rigorous association betwen a FPE and a system of SDEs. If the FPE is written in the form

$$\frac{dP(x)}{dt} = \left[\sum_i \frac{d}{dx_i} A_i(x) + \left[\frac{1}{2}\sum_j \frac{d^2}{dx_i dx_j} D_{ij}(x)\right]\right] P(x) \tag{11}$$

where $A_i(x)$ is the drift vector and $D_{ij}(x)$ is the diffusion matrix, the equivalent SDEs are (Gardiner 1983)

$$dx_i = A_i(x)dt + \sum_j d_{ij} dW_j \tag{12}$$

where the dW_j are the increments of independent Wiener processors and d is a matrix such that $dd^T = D$. There is no unique solution for d; it can for example be multiplied by an orthogonal matrix. One way of obtaining d is to perform a Cholesky decomposition (Sarkar, Satchell and Carmichael, 1986c) on the original diffusion matrix. The diffusion matrix is usually a function of the variables, and so in general d is as well. The SDEs have multiplicative rather than constant noise terms.

Once we have obtained a system of SDEs we can solve it numerically. The standard method is analogous to the Euler method for the solution of ODEs. We can write the solution of the SDE as formally as

$$x_i(t) = x_i(0) + \int_0^t A_i(x(s))ds + \sum_j \int_0^t d_{ij}(x(s))dW_j(s) \tag{13}$$

and if we expand in Taylor series form and truncate at lowest order we obtain

$$\Delta x_i = A_i(x)\Delta t + (\Delta t)^{\frac{1}{2}} \sum_k d_{ik}\, \xi_k \tag{14}$$

where Δt is the simulation time step. The quantities ξ_k are independent Gaussian variables of unit variance; in a computer program they are simulated by calls to a Gaussian distributed pseudo-random number generator.

It is possible to work to higher order in Δt. Usually the equations become unmanageably complex but in some special cases (for example d constant or linear in the variables) these methods are practical (Satchell and Sarkar, 1986).

An important practical point must be noted here. If the equations are simulated with too large a time step they lose stability. But if there are widely different timescales in a problem we may wish to cover many times the fastest timescale in order to see slower phenomena. In this case the equations are termed "stiff", and the simulation is computationally extremely expensive. Usually we can make an adiabatic elimination, but if the short timescales cannot be eliminated (for example near a critical or bifurcation point or when studying chaotic orbits) the statistical accuracy obtainable may be very limited.

In some cases the SDEs become numerically unstable in some regions of phase space. Trajectories which wander into these regions are apt to result in computer overflow, so one normally abandons the trajectory and does not include its contribution into the ensemble. This obviously introduces a bias, but is merely a sympton of the incipient breakdown of the method. This trajectory failure first lead us to discover problems with the positive-P distribution.

5. APPLICATIONS AND EXAMPLES

We will concentrate on two examples drawn from quantum optics. Both can be viewed as special cases of the master equation for N two state atoms interacting with a single mode plane wave field in an optical cavity. There are three types of dissipation in the system. The cavity mode has a decay rate κ (due to the transmission in the cavity mirrors), the excited state atomic population decays by spontaneous emission with rate $\gamma_{\parallel}$ and the atomic polarisation decays with rate $\gamma_{\perp}$. In the absence of collisions

$$\gamma_{\perp} = \gamma_{\parallel}/2$$

The atoms may be pumped (that is some incoherent process raises them to the upper state with rate $\gamma_{\uparrow}$) and there may be an injected field E. The resulting master equation is

$$\dot{\rho} = [H,\rho]/i\hbar + \kappa([a\rho,a^{+}] + [a,\rho a^{+}]) + L_{A}\rho \tag{15}$$

where

$$H = ig\hbar \sum_{\mu=1}^{N} [\exp(-ik\cdot x_{\mu})a^{+}\sigma_{\mu}^{-} - \exp(ik\cdot x)a\sigma_{\mu}^{+}] \tag{16}$$

$$L_{A}\rho = \tfrac{1}{2} \sum_{\mu=1}^{N} \{\gamma_{\uparrow}([\sigma_{\mu}^{+},\rho\sigma_{\mu}^{-}] + [\sigma_{\mu}^{+}\rho,\sigma_{\mu}^{-}]) + \gamma_{\downarrow}([\sigma_{\mu}^{-},\rho\sigma_{\mu}^{+}] + [\sigma_{\mu}^{-}\rho,\sigma_{\mu}^{+}]) + \gamma_{o}([\sigma_{\mu}^{z},\rho\sigma_{\mu}^{z}] + [\sigma_{\mu}^{z}\rho,\sigma_{\mu}^{z}])\} \tag{17}$$

and γ_{o} is the phase decay rate due to collisions, and g is the strength of the coupling of the atom to the cavity field.

5.1 THE LORENZ-HAKEN LASER EQUATIONS

Haken (1975) has shown that the Maxwell-Bloch equations for a single mode homogeneously broadened laser are equivalent, upon rescaling of the variables and constants, to the well known Lorenz equations (Lorenz 1963). These Lorenz equations arose originally in the context of Rayleigh-Benard convection, and show chaos for suitable parameter values. They have been extensively studied; the strange attractor has a complex multi-sheeted structure, with fractal properties (Mandelbrot, 1982). The quantum optical system therefore also shows chaos in the semi-classical Maxwell-Bloch equation description (for pumping parameters, however, that are too strong to be obtainable in most lasers). It is natural to ask about the effects of intrinsic quantum fluctuations on this chaos.

5.1.1 The Wigner Distribution

We consider the case without an injected field first. For consistency we must allow the field (X) and polarisation (Y) variables to be complex here although the noiseless model equivalent to the Lorenz equation takes these real. This leads us to consider a five dimensional phase space. In the absence of noise we recover the standard Lorenz equations on choosing the phase of X and Y to be zero. The SDEs are (Sarkar, Satchell and Carmichael, 1986b)

$$dX = \sigma(Y-X)dt + (2/C)^{\frac{1}{2}}\,\sigma\varepsilon\,(dW_1+idW_2) \tag{18a}$$

$$dY = (XZ-Y)dt + 2\varepsilon(dW_3+idW_4) \tag{18b}$$

$$dZ = -(B(Z-R) + (XY^*+X^*+Y))dt - (BR\varepsilon/2C^2)(\mathrm{Re}(Y)\,dW_3 + \mathrm{Im}(Y)dW_4) + 2\varepsilon\,B^{\frac{1}{2}}\,[2- RZ/2C^2 - (R^2B/16C^4)\,YY^*]^{\frac{1}{2}}\,dW_5 \tag{18c}$$

where the dW_i ($i = 1,\ldots,5$) are the increments of independent Wiener processes and the parameters

$$\sigma = \kappa/\gamma_\perp$$

$$B = (\gamma_\parallel + \gamma_\uparrow)/\gamma_\perp$$

and

$$R = 2C\,\frac{\gamma_\uparrow - \gamma_\parallel}{(\gamma_\uparrow + \gamma_\parallel)}$$

correspond to those of the usual Lorenz model.

We have introduced two additional constants into the equations, C and ε. ε is a measure of the noise strength given by

$$\varepsilon^2 = \frac{C^2B\gamma_\perp}{N\gamma_\uparrow}$$

and C is

$$C = g^2N/\kappa\gamma_\parallel$$

First let us examine the validity of the above truncation approximation of the generalised FPE. N can be found from C.

$$N = \frac{C^2B\gamma_\perp}{\varepsilon^2\gamma_\parallel}$$

But we know that C is given by

$$C = \frac{R(\gamma_{\uparrow} + \gamma_{\parallel})}{2(\gamma_{\uparrow} - \gamma_{\parallel})}$$

Now N has a minimum value when

$$\gamma_{\uparrow} = 5\gamma_{\parallel}$$

and so N the number of atoms must exceed (Sarkar and Satchell, 1986c)

$$\frac{27R^2B}{8\varepsilon^2}$$

For a laser with B = 1.0 (small amount of pressure broadening) and $\sigma = 5.0$ (bad cavity) we might choose R = 20 in the chaotic region. Then $N>1350/\varepsilon^2$.

Even for $\varepsilon = 1.0$, for which we find that the fine structure of the attractor is completely washed out, N is at least 1350. Furthermore, by choosing a larger value of C we can increase N at this constant value of ε, and so we can achieve large N with non-trivial noise in the FPE. In this case the truncation of higher derivatives from the generalised FPE can be made without restricting ourselves solely to the small noise limit.

In the limit of large C values the noise terms effectively simplify. The multiplicative terms are removed, and only the following terms survive:

$2\varepsilon(dW_3+idW_4)$ in the equation for dY

$\varepsilon(8B)^{\frac{1}{2}}dW_5$ in the equation for dZ

On the other hand C must exceed R/2, and this would give the greatest effect for the multiplicative noise term. However the values of the Y and Z variables on the attractor typically of modulus less than R/2, and so the noise term is unlikely to vary by more than a factor of 3.

The standard definition of the fractal dimension in terms of box counting is computationally impractical for multidimensional systems. Instead we use the algorithm of Termonia and Alexadrowicz (1983), which is based on the distribution of spacings between points in phase space from a sampled time series. We have examined the structure of the steady state probability distribution, and find that it shows fractal scaling down to some noise dominated inner scale (Sarkar and Satchell, 1986b). The fractal dimension in the scaling region is greater by one than that for the deterministic Lorenz equations. We identify this increase with phase diffusion of the laser field. At the smallest scales observable with the number of points we used the dimension D_F rises for $\varepsilon = 0.3$ (Table 1).

We presume that it would eventually take the value of five, which we would expect for a trivial fivefold product of noise processes. This rise gives us some additional a posteriori justification for truncating the higher derivatives in the generalised FPE, as it suggests that the probability distribution becomes smooth at small scales.

TABLE 1. The Effect of Inner Scales for $\varepsilon = 0.3$

Range of n	D_F
[1, 10]	3.7
[10, 20]	3.5
[20, 30]	3.38
[50, 60]	3.31
[100, 110]	3.23
[200, 210]	3.18
[490, 500]	3.16

We have also calculated the Lyapunov exponents for the process. The standard definition of the exponents is not applicable to a non-deterministic system, as it considers the evolution of two infinitesimally separated trajectories in the limit that times goes to infinity, i.e.

$$\lambda = \lim_{t\to\infty} \frac{1}{t} \ln \left[\frac{\Delta x(t)}{\Delta x(0)}\right]$$

where λ is the exponent and Δx is the separation of the two trajectories.

In a probabilistic system we no longer have trajectories, but instead the evolution of probability distributions. Each of these will spread out to cover the attractor and so the concept of their separation is meaningless. We cannot apply this definition to the noisy case. Instead we follow the method suggested by Eckmann and Ruelle (1985) for estimating the exponents from experimental time series. Their method requires the definition of the tangent space equations for a trajectory, and then integrating these equations along a typical trajectory that covers the attractor. The tangent equations define a linear map between initial infinitesimal deviations from the starting point and those at the final point. Diagonalising this map gives the exponents directly. In practice we cannot integrate these equations indefinitely for numerical reasons, but we build the map up as the product of stroboscopic maps for sections of the trajectory. In the deterministic case this gives the same results as the more usual approach, but we can easily extend it to the noisy case (Sarkar and Satchell, 1987a). The tangent space equations describe the local rate of divergence of convergence of the centres of peaked probability distributions. We find the geometric average of these rates for the attractor as a whole by following a single stochastic trajectory, and applying the same prescription.

As we are dealing with a five dimensional phase space we have five exponents. One is positive, and identical (for small noise) with the value expected for the positive exponent in the deterministic three variable Lorenz equations. There are two zero exponents, which we can identify with the time translation and phase rotation symmetries of the problem.

The standard equations show the time translation symmetry, but the phase rotation symmetry only arises when X and Y are complex. Finally there are two negative exponents, one of which can be identified with the negative exponent of the standard equations. The new negative exponent is associated with the relaxation of phase differences between the X and Y variables. Associated with these exponents is a Lyapunov dimension and this retains a value close to its deterministic one. The Lyapunov dimension "knows" about the additional phase diffusion degree of freedom even though the equations do not explore these in the absence of noise. In this case we find that the fractal and Lyapunov dimensions only take on similar values in the presence of noise. (For axiom A systems the Lyapunov and fractal dimensions are equal (Eckmann and Ruelle, 1985). The deterministic laser equations are not an axiom A system.)

The phase symmetry of the system can be broken by adding an injected field. This changes one of the zero exponents to a negative one (the system now has a preferred phase to which it is attracted) and the fractal dimension is no longer increased by one over its deterministic value. In fact we no longer find a proper scaling region, as the probability distribution in the phase directions is small compared with the total size of the attractor but large compared with the other noise dominated inner scales.

The evolution towards a steady state probability distribution for the Lorenz system provides a fascinating example of the ideas of algorithmic complexity (Ford, 1986) in quantum chaos. Ford has provided arguments that Hamiltonian quantum mechanical systems are 'computable', while deterministically chaotic systems are 'non-computable'. Although we find positive Lyapunov exponents this does not give algorithmic complexity. The presence of noise in the system and the strongly contracting (dissipative) flow of the Lorenz equations mean that the system soon loses all memory of its initial conditions. The probability distribution soon spreads over all of the attractor, and rapidly approaches the steady state. From this time on, further computation is unneeded. The strange attracting set of the Lorenz equations is fractal: it contains structure on arbitrarily small scales. The laser equations show fractal properties down to some inner scale, but are smooth over small enough distances. The amount of information required to describe the deterministic strange attractor increases without limit as we go to smaller scales, but for the laser equations with their intrinsic fluctuations only a finite amount of information is required to describe the large time probability distribution. Hence dissipative quantum systems are 'computable'.

5.1.2 The Positive-P Distribution

If we follow the standard prescription for the construction of the positive-P distribution for the laser we obtain a ten dimensional phase space with additional non-physical degrees of freedom. The SDEs are (Sarkar, Satchell and Carmichael, 1986c).

$$dX_1 = \sigma(Y_1-X_1)dt \tag{19a}$$

$$dX_2 = \sigma(Y_2-X_2)dt \tag{19b}$$

$$dY_1 = (-Y_1+X_1Z)dt + 2^{3/2}\,\varepsilon \sum_{\nu=1}^{3} d_{1\nu}dW_\nu \tag{19c}$$

$$dY_2 = (-Y_2+X_2Z)dt + 2^{3/2}\,\varepsilon \sum_{\nu=1}^{3} d_{3\nu}dW_\nu \tag{19d}$$

$$dZ = B(R-Z)dt - (Y_1X_2+Y_2X_1)dt + 2\varepsilon \sum_{\nu=1}^{3} d_{2\nu}dW_\nu \tag{19e}$$

where $dd^T = D$; the symmetric diffusion matrix is defined by

$$D = \begin{pmatrix} \frac{4X_1Y_1}{C} & \frac{-2b}{C}Y_1\left[1+\frac{R}{2C}\right] & \frac{4}{C}(Z+2C) \\ & \frac{2B}{C}\left[4C-\frac{RZ}{C}-\frac{2}{B}(X_1Y_2+X_2Y_1)\right] & \frac{-2b}{C}Y_1\left[1+\frac{R}{2C}\right] \\ & & \frac{4X_2Y_2}{C} \end{pmatrix} \tag{20}$$

(The unspecified elements of D are determined by symmetry.)

First we consider the deterministic behaviour of this system of equations. The steady states of Eqns. (19a-19e) for $R > 1$ are given by

$$X_1 = Y_1 = X_1^{(0)} = |X_1^{(0)}|\,e^{i\phi}$$
$$X_2 = Y_2 = X_2^{(0)} = |X_2^{(0)}|\,e^{-i\phi}$$
$$|X_1^{(0)}||X_2^{(0)}| = B(R-1)/2$$

For $R < 1$ we find

$$X_1 = Y_1 = X_1^{(0)} = |X_1^{(0)}|\,e^{i\phi}$$
$$X_2 = Y_2 = X_2^{(0)} = -|X_2^{(0)}|\,e^{-i\phi}$$
$$|X_1^{(0)}||X_2^{(0)}| = B(R-1)/2$$

and in addition there is a state

$$X_1 = X_2 = Y_1 = Y_2 = 0$$
$$Z = R$$

Conjugacy for steady states would require relations such as $X_1^* = X_2$.

The steady states for which conjugacy does not hold are unphysical (in the classical limit). A linearised stability analysis can be performed for all of these states. The eigenvalues occur in pairs (because of the symmetry with which the additional degrees of freedom appear) and satisfy

$$\lambda(\lambda+\sigma+1)\,[\lambda(\lambda+\sigma+1)(\lambda+B) + B(R-1)(\lambda+2\sigma)] = 0$$

The cubic in the brackets appears in the analysis of the real Lorenz equations. Consequently there is a subcritical Hopf bifurcation to chaos for our system at the same value of r (> 1) for all steady states. There is a difference from the standard Lorenz equations, however. When the fixed points lose stability there are two eigenvectors associated with the positive eigenvalues. This is not consistent with the behaviour of the standard Lorenz equations. Furthermore the eigenvectors point in unfortunate directions; when the standard equations show chaos (i.e. when the real parts of one pair of eigenvalues are positive) there is tendency for the imaginary part of the inversion variable Z to grow. This leads to a runaway breakdown of conjugacy.

The phenomenon of the doubling of the number of unstable directions of fixed points is quite general. It follows from the symmetry with which the additional degrees of freedom appear (Sarkar, Satchell and Carmichael, 1986d).

If we take an initial condition that is conjugate (i.e. physical), the deterministic solutions do preserve conjugacy. If however the system is perturbed from this initial condition the conjugacy is not maintained in the chaotic region. In fact the trajectories move on to a limit cycle rather than the standard chaotic attractor (see Figure 1). We can examine

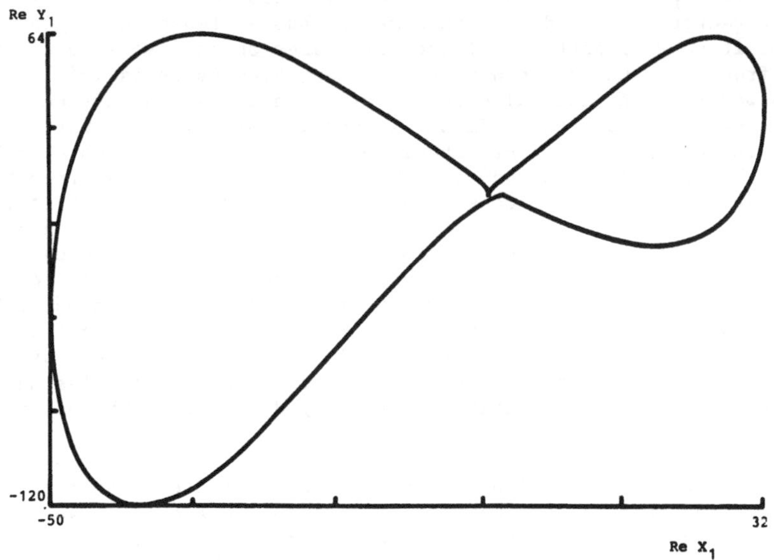

FIGURE 1.
Limit cycle attractor for the deterministic positive P equations for a laser in the chaotic regime. R = 28, σ = 5, b = 8/3.

this by considering the evolution of a system of equations with a small perturbation from conjugacy. We consider a reduced set of equations that are valid on the attractor

$$dX/dt = -\sigma(X-Y)$$

$$dY/dt = -Y + XZ$$

$$dZ/dt = -B(Z-R) - XY$$

where X, Y and Z are in general complex. Conjugacy would require these reduced variables to remain real. We take an initial condition

$$X = X_r + i\delta X$$

$$Y = Y_r + i\delta Y$$

$$Z = Z_r + i\delta Z$$

The growth or decay of the small deviations δX, δY and δZ is given by the eigenvalues of the matrix

$$\begin{pmatrix} -\sigma & \sigma & 0 \\ Z_r & -1 & X_r \\ -Y_r & -X_r & -B \end{pmatrix}$$

where X_r, Y_r, Z_r are the values from the evolution of a trajectory on the normal attractor. We find that this matrix has at least one eigenvalue with a positive real part for almost all places on the attractor. It appears from this that the non-physical quantities (such as Im(Z)) will grow. However if the initial ensemble of trajectories is physical, it will have symmetry such that Im(Z) is zero. For any trajectory which deviates in one direction there should be a mirror image which cancels its contribution. The tendency for the solution to grow away from the standard attractor also has consequences for the physical variables. The limit cycle in Figure 1 has a mirror image which allows the non-physical variables to average to zero for an ensemble. This does not change the fact that we should find chaos and not a limit cycle for these parameter values. Furthermore physical variables (such as Re(Z)) take on values that lie well outside the range of the Lorenz attractor. If the deterministic solutions move in this non-physical manner averaging cannot recover sensible behaviour. On the other hand in the lasing case, where the steady states are stable, averaging does work. The non-physical steady states form a stable continuum.

We have examined the evolution of the equations for the noisy case numerically. We find that in the lasing region the noise averages over the non-physical steady states, and the expectation values of operators have plausible values. In the chaotic region we find that the noise soon leads to an irrecoverable loss of conjugacy and (frequently) divergence of the trajectories. This occurs for levels of noise for which the Wigner SDEs are entirely satisfactory.

We are forced to conclude that although the positive-P distribution is satisfactory in the lasing and sub-lasing regions, above the chaos threshold the additional dimensions allow extra instabilities. These instabilities are sufficiently severe to drastically alter the nature of the solution from that which is expected. We now speculate on the cause of the difficulty. It is possible that although the distribution explicitly constructed from the density matrix satisfies the FPE, it is not the only distribution that does so. The dynamical instability apparently leads the correct physical distribution to evolve into a non-physical one.

The validity of this distribution has not been proven for the cases (such as ours) where atomic operators occur. In a more recent study Dörfle and Schenzle (1986) have found an example of similar breakdown in a case where only bosonic (field) operators were used. Although the positive-P can be constructed, the solutions are unstable and the simulation also fails.

5.2 Absorptive Optical Bistability

Optical bistability has been extensively studied both experimentally and theoretically. Good agreement has been reached between theory and experiment for steady state quantities, but the quantum statistical predictions of the theory are only beginning to be explored in experiment. Our interest here in optical bistability arises because it is one of the simplest physical systems to show a first order phase transition.

This system can show one or two stable fixed points for X, the scaled intracavity field, depending on the value of C, the co-operativity parameter, and Y, the scaled input field. These are given by

$$C = \alpha L/2T \qquad X = E/E_s \qquad Y = E_i(E_s T^{\frac{1}{2}})$$

$$E_s = h(\gamma_\perp \gamma_\parallel)^{\frac{1}{2}}/d$$

where αL is the single pass absorption in the medium, T is the transmission of the input and output mirrors and d is the modulus of the dipole matrix element. The deterministic steady states satisfy

$$Y = X\left(1 + \frac{2C}{1+X^2}\right)$$

When $C > 4$ there are two stable and one unstable fixed points for suitable values of input field. For $C < 4$ there is only ever one fixed point. When $C = 4$ and $Y = 3\sqrt{3}$ we find a critical case, and there is enhanced sensitivity to fluctuations. We will use this as a test case for comparing the solutions obtained by quasi-probability methods with the exact solution of the master equation.

We will consider a very simple case; N two state atoms resonantly coupled to a single mode cavity. If the relaxation rate for the cavity mode is much more rapid than the atomic relaxation we can adiabatically eliminate the field operators and express them in terms of atomic operators. This gives us a simple ME with a finite basis of states. We have been able to study the solutions of the ME with up to 45 atoms, and this allows us to overlap (just) with the region for which we can study the Ito SDEs without excessive trajectory failure (Sarkar and Satchell, 1987b).

We have considered two distribution functions here. The Wigner distribution yields a FPE without positive definite diffusion. It is however positive definite at the fixed points, so linearised analysis is always possible. The positive-P distribution gives a FPE with positive definite diffusion but suffers from a milder form of the pathology seen above. This leads to some important practical difficulties in its application.

It must be emphasised that the truncation involved in obtaining the FPE is completely different in the two cases and so there is no a priori reason to believe that a nonlinear analysis of the FPEs should agree.

5.2.1 Master Equation (ME) Solution

The master equation can be converted to a system of c-number equations by choosing a suitable basis. The bad cavity ME only involves the atomic operators since the field operators have been adiabatically eliminated.

$$\begin{aligned}\dot{\rho} &= L\rho \\ &= -i\hat{\Omega}\sum_{\mu=1}^{N}[\sigma_\mu^+ + \sigma_\mu^-, \rho] + \frac{g^2}{\kappa}\sum_{\mu=1}^{N}(2\,\sigma_\mu^-\rho\sigma_\nu^+ - \sigma_\mu^+\sigma_\nu^-\rho - \rho\sigma_\mu^+\sigma_\nu^-) \\ &\quad + \frac{\gamma_\parallel}{2}\sum_{\mu=1}^{N}(2\,\sigma_\mu^-\rho\sigma_\mu^+ - \sigma_\mu^+\sigma_\mu^-\rho - \rho\sigma_\mu^+\sigma_\mu^-) \\ &\quad + \tfrac{1}{2}\gamma_0\sum_{\mu=1}^{N}(2\,\sigma_\mu^3\rho\sigma_\mu^3 - \tfrac{1}{2}\rho)\end{aligned}$$

$\hat{\Omega}$ is proportional to the classical driving field.

Here L represents the Liouvillian (super-operator) which describes the time evolution of the density matrix. This yields a problem with a finite basis. We can for example choose the direct product of the single atom basis vectors. This rapidly becomes unwieldy since for N atoms the basis has 2^N elements, and the density matrix 2^{2N}. However most of these elements are redundant, as all atoms appear in the problem in the same way. To be more precise the ME has symmetry under any permutation of atom labels. We can make use of this symmetry to find a smaller subset of distinct density matrix elements. Having chosen such a set the problem reduces to a large system of coupled linear ODEs. This allows us to solve the master equation for as many as 45 atoms for which we have 35,000 coupled equations. We would have been unable to treat more than 6 atoms using the direct product basis. Larger systems will be possible with improved computers. We evaluate three quantities for comparison with the FPE: $\langle X\rangle$, $\langle XX^*\rangle$ and the normalised 2nd moment of the intensity distribution $g^2(0)$.

The ME is linear operator equation and always has a unique steady state solution. Fluctuations induce transitions between the stable states of the deterministic system. The steady state solutions of the ME represent an equilibrium statistical mixture of the two states. Optical bistability, like all phase transitions, is an emergent phenomenon which appears in the limit of large system size. One quantum mechanical signature of bistability is a long relaxation time, longer than any of the natural relaxation times in the system corresponding to a transition time between bistable (or metastable) states. We have found evidence for the existence of slow relaxation processes by directly diagonalising the Liouvillian super-operator. This has been only practical so far for systems of up to twelve atoms (Sarkar and Satchell, 1987c).

5.2.2 Positive-P SDEs

The bad cavity SDEs are (Carmichael et al, 1983)

$$dZ_1 = -\nu\,(Z_2 - Z_3(4CZ_2+Y))\,d\tau + \sqrt{\frac{\nu}{N}}\,\sqrt{Z_1(4CZ_1+Y)}\,dW_1 \qquad (23a)$$

$$dZ_2 = -\nu\,(Z_2 - Z_3(4CZ_2+Y))\,d\tau + \sqrt{\frac{\nu}{N}}\,\frac{(2\nu-1)(\tfrac{1}{2}+z_3)}{\sqrt{Z_1(4CZ_1+Y)}}\,dW_1 + \sqrt{\frac{\nu}{N}}\sqrt{Z_2(4CZ_2+Y - \frac{(2\nu-1)^2(\tfrac{1}{2}+Z_3)^2}{(4CZ_1+Y)Z_1}}\,dW_2 \qquad (23b)$$

$$dZ_3 = -(\tfrac{1}{2} + Z_3 + \tfrac{1}{2}(Y(Z_1+Z_2) + 8CZ_1Z_2))\,d\tau + \frac{1}{\sqrt{N}}\,(\tfrac{1}{2} + Z_3 - \tfrac{1}{2}(Y(Z_1+Z_2) + 8CZ_1Z_2))^{\frac{1}{2}}\,dW_3 \qquad (23c)$$

where Z_1, Z_2 and Z_3 are complex variables related by scalings to

$$\sum_\mu \sigma_\mu^- \;,\quad \sum_\mu \sigma_\mu^+ \quad \text{and} \quad \sum_\mu \sigma_\mu^3 \;.$$

ν is defined to be $\gamma_\perp/\gamma_\parallel$. The field variable is $(4C(Z_1 + iZ_2) + Y)$.

We note that the inversion variable is now complex, although it must remain real in the mean for a physically acceptable solution. In a similar way the two polarisation variables are not constrained to be conjugates of one another.

We first consider the simple case of $C = 2$ and $Y = \sqrt{2}$. There is a single stable fixed point here, with relatively low saturation of the atoms. Not surprisingly we find that the results of linearised analysis are in good agreement with the ME, except for the very smallest numbers of atoms. If we use the SDEs to perform a nonlinear analysis we find that many trajectories leave the physical (or 'classicially' allowed) region. It is essential to include these, otherwise we introduce a severe statistical bias into the ensemble. The effect of rejecting trajectories which leave the physical region is shown in Figures 2, 3, 4 and 5. When we do retain all the trajectories we find an interesting result; the linearised analysis agrees marginally better with the ME than the full FPE as shown in Figure 2 for the expectation value of the intensity I. The overall trend (from Figure 3) for the quantity $N(1-g^2(0))$ is to drop at small values of N for the ME. On the other hand it rises slightly for the FPE, while it is of course constant for linearised analysis. The nonlinear analysis does not improve the agreement in this case, presumably because the processes which are nonlinear in the fluctuations are no more important than those which have been removed by the truncation of the higher derivatives.

The drift terms in the SDEs show an additional set of fixed points for cases in which bistability is found. These points are non-physical in the sense that they are non-conjugate. The spacing of these points from the normal solutions depends on the size of the hysteresis loop. When the spacing is large and the noise small trajectories remain in the vicinity

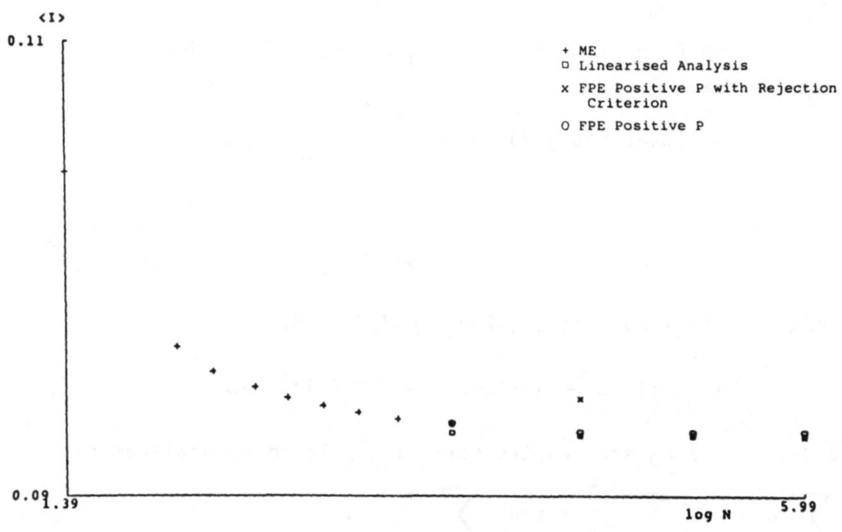

FIGURE 2.
Expectation value of scaled intensity as a function of number of atoms. $Y = \sqrt{2}$, $C = 2$.

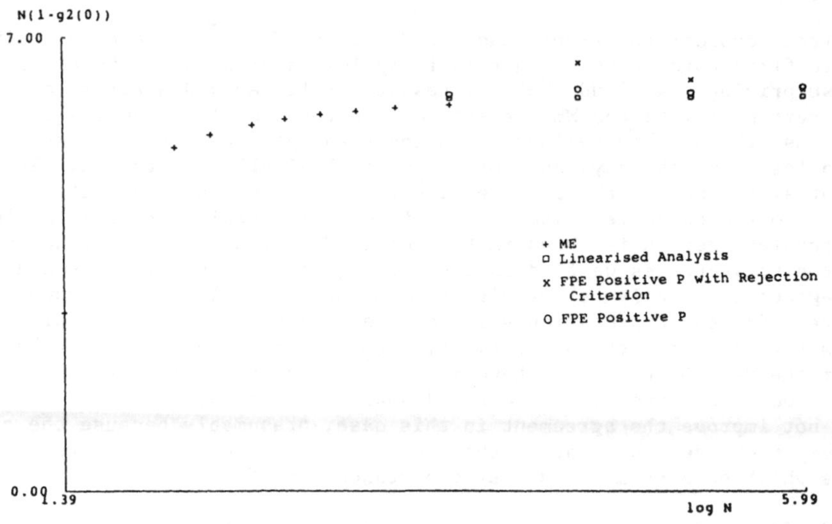

FIGURE 3.
$N(1-g^2(0))$ as a function of number of atoms. $Y = \sqrt{2}$, $C = 2$.

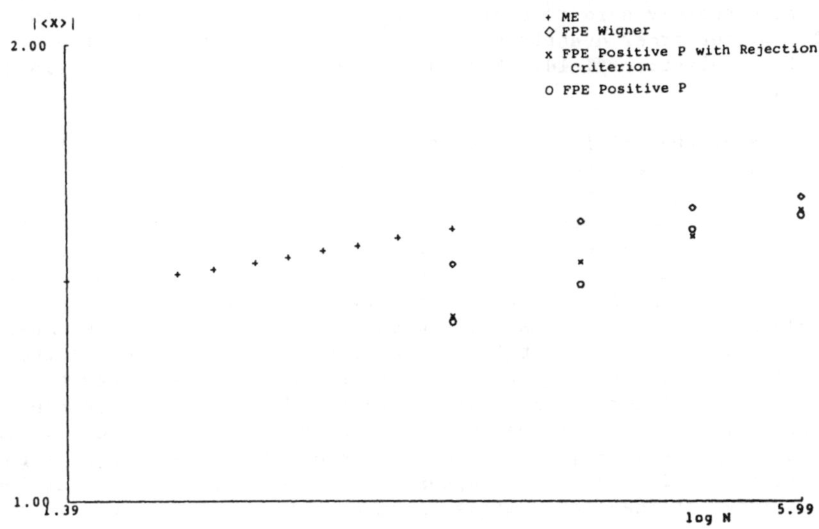

FIGURE 4.
Expectation value of intracavity field for C = 4, Y = $3\sqrt{3}$.

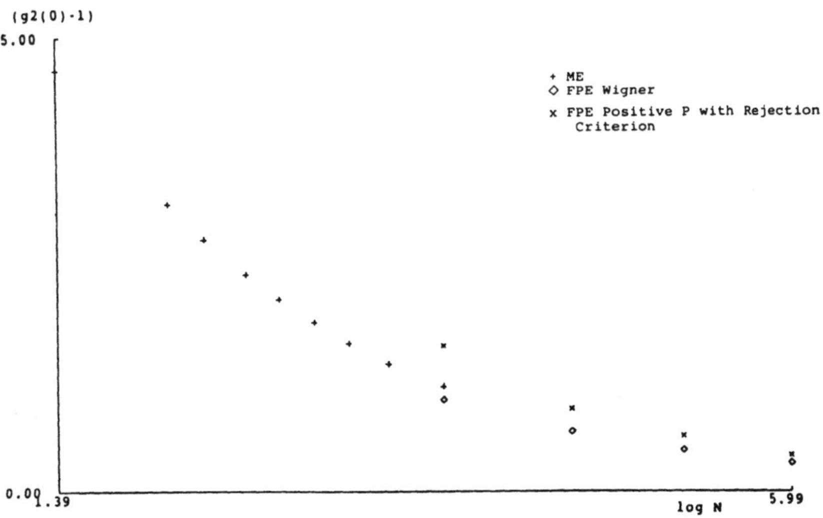

FIGURE 5.
$(g^2(0)-1)$ as a function of number of atoms. C = 4, Y = $3\sqrt{3}$.

of the physical fixed points. As the noise is increased, or the additional solutions approach the physical ones, trajectories begin to visit these points in phase space. This results in rare but large transients, which are extremely hard to average away. This is particularly troublesome for higher order quantities like $g^2(0)$, which depends on the 4th power of the electric field. Eventually, at $C = 4$, all the solutions coalesce.

In the case of the critical point at $C = 4$ and $Y = 3\sqrt{3}$ no linearised analysis is possible, and we compare the full nonlinear FPE with the ME. There are two eigenvectors with zero eigenvalues for the fixed point, and the trajectories penetrate deep into the non-physical region. In doing so they experience large transients which give rise to short bursts of very large amplitude. These bursts effect single trajectories, and are very hard to average away. The qualitative trend for $\langle X \rangle$ in Figure 4 and $(g^2(0)-1)$ in Figure 5 is consistent with the ME solution but as expected the large fluctuations result in much poorer numerical agreement than for the previous case. The qualitative trend for the mean intracavity scaled intensity I in Figure 6 is to drop with N. This is in direct contrast with the ME solution. This may be due to a change in the effective value of the driving field (Y), as the dependence of $\langle X^2 \rangle$ on N changes sign for values of Y in this region. It appears that the additional symmetries of the positive-P are causing a milder form of the pathology found in the laser.

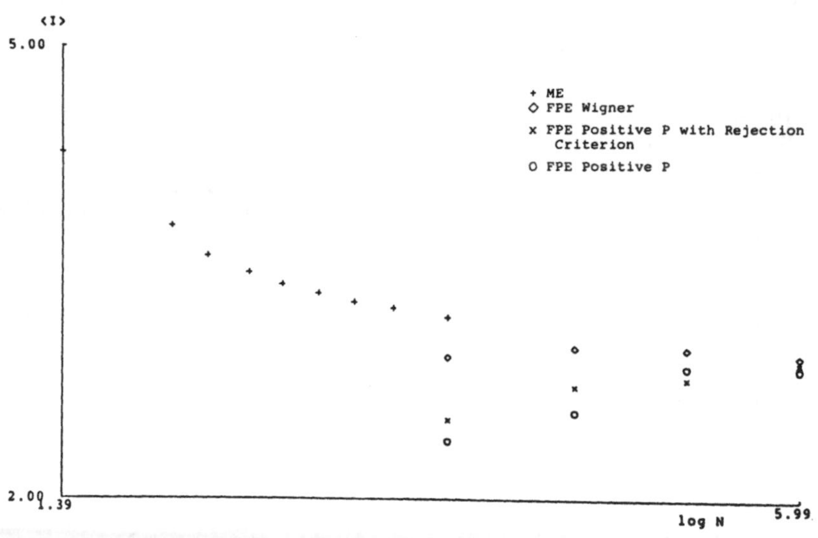

FIGURE 6.
Dependence of mean scaled intensity on number of atoms. $C = 4$, $Y = 3\sqrt{3}$.

5.2.3 Wigner Distribution

The SDEs for the polarisation and inversion variables are (Sarkar, Satchell and Carmichael (1986b)

$$dZ_1 = -\nu(Z_1 - (Y+4CZ_1)Z_3)d + \nu\sqrt{\frac{2}{N}}\,dW_1 + \nu\sqrt{\frac{4C}{N}}\,Z_3\,dW_4 \quad (24a)$$

$$dZ_2 = -(Z_2 - 4CZ_2Z_3)d + \nu\sqrt{\frac{4C}{N}}\,Z_3\,dW_5 + \nu\sqrt{\frac{2}{N}}\,dW_2 \quad (24b)$$

$$dZ_3 = -(Z_3+\tfrac{1}{2}+YZ_1+4C(Z_1^2+Z_2^2)) - \sqrt{\frac{4C}{N}}\,(Z_1dW_4 + Z_2dW_5)$$

$$+ \sqrt{\frac{1}{2N\nu^2}}\,Z_1\,dW_1 + \sqrt{\frac{1}{2N\nu^2}}\,Z_2\,dW_2 \quad (24c)$$

$$+ \sqrt{\frac{1}{N}\,(Z_3+\tfrac{1}{2}) - \frac{1}{2\nu^2N}\,(Z_1^2+Z_2^2)}\;dW_3$$

where Z_1, Z_2 and Z_3 are real variables which are related by scaling factors to the real and imaginary parts of the polarisation and the inversion (which is real). (The field variable X can be expressed in terms of Z_1, Z_2 and Z_3.)

We can easily show that the diffusion is always positive definite at a fixed point. This means that linearised analysis around stable fixed points is always valid. However if we attempt to go beyond this to a nonlinear analysis of the FPE we soon find that the Ito equations may fail some cases. For $C = 2$ and $Y = \sqrt{2}$ there is a single stable fixed point. The linearised analysis about this agrees surprisingly well with the ME, but an attempt to perform a nonlinear analysis by means of SDEs soon fails as the trajectories enter a region of non-positive definite diffusion. When $C = 4$ and $Y = 3\sqrt{3}$ the system has a fixed point of neutral stability. In this case no linearised analysis is possible. The results of a nonlinear analysis show reasonable agreement with the ME. There were no trajectories which entired the region of non-positive definite diffusion for large N (of order 100). The qualitative trends for $\langle X \rangle$ and $g^2(0)$ are in reasonable agreement with those for the ME, while $\langle I \rangle$ is approximately independent of N. Overall the Wigner FPE solution is in better agreement with the ME in this case.

The relationship between the rate of fluctuation induced transitions from the master equation and that from an FPE is still an open question. In these non-potential multi-dimensional cases there are no analytic tools for calculating the tunnelling rate. However one can in principle estimate the rate from the SDEs. This is only possible for reasonably large values of N (say greater than 50) because trajectory failure becomes excessive for smaller values. Such a comparison must await the availability of computers which will allow us to diagonalise very large matrices. For example the Louvillian for a system with 50 atoms is a 23426 x 23426 non-Hermitian complex matrix, and this lies well beyond the current state of the art. Because the rates depend critically on the shape of the tails of the probability distribution we expect that this may well be very sensitive to the details of the truncation used.

6. CONCLUSION

The power of quasi-probability methods, when justified, has been demonstrated dramatically for the Lorenz equations. We were able to examine comprehensively both the static and dynamic properties of the solution through the Wigner function. The example of absorptive bistability showed a less rosy picture. The FPE may not have positive definite diffusion for physically relevant examples. This brings us to the heart of the problem. When a well behaved FPE can be obtained, it is a powerful calculational tool, but there are many physically important systems for which this is not possible. The positive-P distribution always has positive definite diffusion, but suffers from non-physical instabilities when the system shows chaos or is at a critical point.

The present state of the art is somewhat frustrating. In those cases where it is usable the Wigner distribution works very nicely. Unfortunately many systems do not show positive definite diffusion. Those that do only reveal classical noise processes (although these may be of quantum mechanical origin). When the Wigner distribution is not usable there are few practical options. The Glauber P distribution is frequently highly singular; again this is expected whenever non-classical noise processes occur. The pathology of the positive-P may mean that it is unworkable except for systems with simple and well isolated fixed points. There are a number of other more rarely used distribution functions (Gardiner 1983). Some of these (the complex-P and R functions) are not real valued and hence cannot give rise to SDEs. Others like the Q distribution (the diagonal part of the density matrix in the coherent state basis) suffer from non-positive definite diffusion for some cases of interest. There is as yet no known distribution for handling non-analytic (ie numerical) work with instabilities and nonclassical states. This arguably, is the most important theoretical problem facing the development of the study of quantum effects on instabilities in dissipative systems.

REFERENCES

Berry, M.V., 1983, "Chaotic Behaviour of Deterministic Systems", Eds. G. Iooss, R.H.G. Helleman and R. Stora, North Holland, Amsterdam, pp 171-271.

Carmichael, H.J., 1986, Quantum-Statistical Methods in Quantum Optics, Springer, Berlin.

Carmichael, H.J., Satchell, J.S. and Sarkar, S., 1986, Phys. Rev. A34 3166.

Carmichael, H.J., Walls, D.F., Drummond, P.D. and Hassan, S-S., 1983, Phys. Rev. A27 3112.

Casati, G., 1985, Ed., Chaotic Behavior in Quantum Systems, Plenum, New York.

Dörfle, M. and Schenzle, A., 1986, Z. Phys. 65 113.

Drummond, P.D. and Gardiner, C.W., 1980, J. Phys. A13 2353.

Eckmann, J-P and Ruelle, D., 1985, Rev. Mod. Phys. 57 617.

Ford, J., 1986, Chaotic Dynamics and Fractals, Eds. M.F. Barnsley and S.G. Demko, Academic Press, Orlando, pp 1-52.

Gardiner, C.W., 1983, Handbook of Stochastic Methods, Springer-Verlag, Berlin.

Gronchi, M. and Lugiato, L.A., 1978, Lett. Nuovo Cimento 23 593.

Haken, H., 1970, Laser Theory, Encycl. of Phys. XXV/2C, Springer-Verlag, Berlin.

Haken, H., 1975, Phys. Lett. A53, 77.

Lichtenberg, A.J. and Lieberman, M.A., 1983, Regular and Stochastic Motion, Springer-Verlag, New York.

Lorenz, E.N., 1963, J. Atmos. Sci. 20 130.

Lugiato, L.A., Casagrande, F. and Pizzuto, L., 1982, Phys. Rev. A26 3843.

Mandelbrot, B.B., 1982, The Fractal Geometry of Nature, Freeman, San Francisco.

Pike, E.R. and Sarkar, S., 1986, Eds., Frontiers on Quantum Optics, Adam Hilger, Bristol.

Sarkar, S., 1986a, Ed., Nonlinear Phenomena and Chaos, Adam Hilger, Bristol.

Sarkar, S., Satchell, J.S. and Carmichael, H.J., 1986b, J. Phys. A19 2751.

Sarkar, S., Satchell, J.S. and Carmichael, H.J., 1986c, J. Phys. A19 2765.

Sarkar, S. and Satchell, J.S., 1987a, Phys. Rev. A, 35 398.

Sarkar, S. and Satchell, J.S., 1987b, Solution of Master Equations for Small Bistable Systems, J. Phys. A (to be published).

Sarkar, S. and Satchell, J.S., 1987c, Optical Bistability with Small Numbers of Atoms, Europhysics Letters (to be published).

Satchell, J.S. and Sarkar, S., 1986, J. Phys. A19 2737.

Termonia, Y. and Alexandrowicz, Z., 1983, Phys. Rev. Lett. 51 1265.

van Kampen, N.G., 1981, Stochastic Processes in Physics and Chemistry, North-Holland, Amsterdam.

INDEX

Milton Keynes UK
Ingram Content Group UK Ltd.
UKHW040107071024
449327UK00019B/881